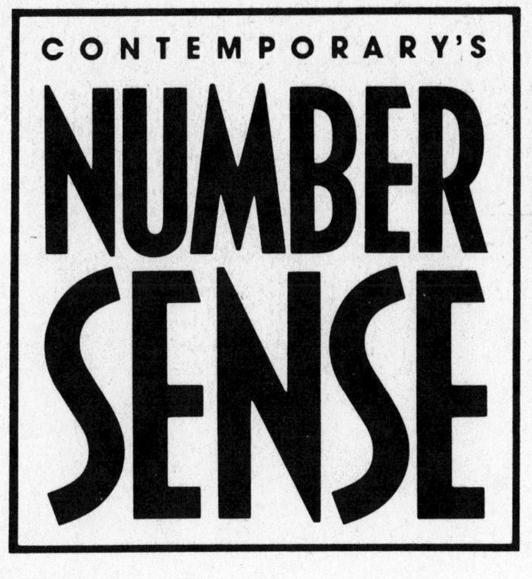

Discovering Basic Math Concepts

Answer Key

Allan D. Suter

Project Editors
Sarah Conroy
Kathy Osmus

CONTEMPORARY
BOOKS
CHICAGO

Copyright © 1990 by Allan D. Suter
All rights reserved

No part of this publication may be reproduced, stored in a retrieval system, or transmitted in any form or by any means, without the prior written permission of the publisher.

Published by Contemporary Books, Inc.
180 North Michigan Avenue, Chicago, Illinois 60601
Manufactured in the United States of America
International Standard Book Number: 0-8092-4191-9

Published simultaneously in Canada by
Beaverbooks, Ltd.
195 Allstate Parkway
Valleywood Business Park
Markham, Ontario L3R 4T8
Canada

Editorial Director
Caren Van Slyke

Editorial
Sarah Conroy
Kathy Osmus
Seija Suter
Steve Miller
Laura Larson
Cliff Wirt
Lynn McEwan
Jane Lichty
Lisa Dillman
Scott Guttman
Robin O'Connor

Editorial Production Manager
Norma Fioretti

Cover Design
Lois Koehler

Art & Production
Arvid Carlson

Typography
Impressions, Inc.
Madison, Wisconsin

Dedicated to our friend, Pat Reid

Contents

CHAPTER 1: PLACE VALUE

Page 1: Hundreds

1. a) 8
   b) 9
2. a) 7
   b) 2
3. a) 4
   b) 0
   c) 5
4. a) 9
   b) 1
   c) 3
5. a) 3
   b) 0
   c) 0

Page 2: Thousands

1. a) 6
   b) 0
   c) 9
   d) 3
2. a) 4
   b) 3
   c) 6
   d) 1
   e) 2
3. a) 2
   b) 6
   c) 4
   d) 8
   e) 0
   f) 2

Page 3: Place-Value Readiness

Place-Value Chart

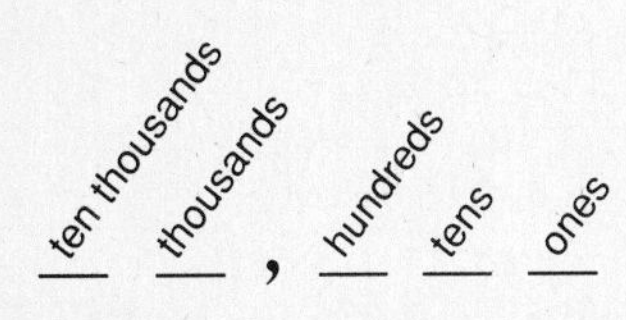

1. thousands
2. tens
3. hundreds
4. ten thousands
5. ones
6. 5
7. 3
8. 8
9. 7
10. 2

Page 4: Number Sense

	BILLIONS			MILLIONS			THOUSANDS			ONES		
	HUNDRED BILLIONS	TEN BILLIONS	BILLIONS	HUNDRED MILLIONS	TEN MILLIONS	MILLIONS	HUNDRED THOUSANDS	TEN THOUSANDS	THOUSANDS	HUNDREDS	TENS	ONES
1. 689 →										6	8	9
2. 6,093 →									6	0	9	3
3. 149,508 →							1	4	9	5	0	8
4. 23,154 →								2	3	1	5	4
5. 5,904,165 →						5	9	0	4	1	6	5
6. 196 →										1	9	6
7. 62,949,150 →					6	2	9	4	9	1	5	0
8. 3,992 →									3	9	9	2
9. 649,132,001 →				6	4	9	1	3	2	0	0	1
10. 326,194,600,109 →	3	2	6	1	9	4	6	0	0	1	0	9

Page 5: Commas

A. 346
B. 12,346
C. 21,612,346
D. 7,421,612,346

1. 7,349
2. 21,496
3. 14,948
4. 964,821
5. 1,436,849
6. 29,743,268
7. 1,413,962
8. 217,924
9. 1,439
10. 341
11. 11,682,371
12. 46,218

Page 6: Words to Numbers

1. 509
2. 703
3. 2,116
4. 5,008
5. 6,843
6. 74,904
7. 7,004
8. 317,052,000
9. 29,704,500,396
10. 9,015
11. 20,045
12. 30,436,165

Page 7: Expanded Forms

1. $(7 \times 10) + (9 \times 1)$
2. $(1 \times 100) + (9 \times 10) + (6 \times 1)$
   $100 + 90 + 6$

Page 7: Expanded Forms (continued)

3. (6×1,000) + (0×100) + (9×10) + (3×1)
   6,000 + 0 + 90 + 3
4. (9×10,000) + (4×1,000) + (0×100) + (8×10) + (9×1)
   90,000 + 4,000 + 0 + 80 + 9
5. (5×1,000) + (1×100) + (4×10) + (6×1)
   5,000 + 100 + 40 + 6

Page 8: Ordering Numbers

1. 83 115 380 651 4,022
2. 1,094 802 290 136 75
3. 158
4. 237
5. 249
6. 7,420
7. 8,532
8. 9,610
9. 75,430
10. 95,431

Page 9: Comparing Numbers

1. 48
2. 513
3. 9,048
4. 55,246
5. <
6. >
7. >
8. =
9. >
10. >
11. >
12. >

CHAPTER 2: ADDITION

Page 10: Meaning of Addition

A. 2 + 4 = 6
B. 6
1. 4 + 5 = 9
2. 5 + 7 = 12
3. 3 + 5 = 8
4. 5 + 8 = 13

Page 11: Addition Facts

+	+0	+1	+2	+3	+4	+5	+6	+7	+8	+9
1	1	2	3	4	5	6	7	8	9	10
2	2	3	4	5	6	7	8	9	10	11
3	3	4	5	6	7	8	9	10	11	12
4	4	5	6	7	8	9	10	11	12	13
5	5	6	7	8	9	10	11	12	13	14
6	6	7	8	9	10	11	12	13	14	15
7	7	8	9	10	11	12	13	14	15	16
8	8	9	10	11	12	13	14	15	16	17
9	9	10	11	12	13	14	15	16	17	18

Page 12: Practice Helps

1. 14
2. 11
3. 15
4. 11
5. 13
6. 16
7. 15
8. 13
9. 12
10. 16
11. 13
12. 17
13. 13
14. 6
15. 14
16. 15
17. 13
18. 10
19. 11
20. 16
21. 17
22. 16
23. 10
24. 15
25. 18
26. 17
27. 13
28. 16
29. 12
30. 17
31. 14
32. 14
33. 12
34. 12
35. 15
36. 7
37. 16
38. 13
39. 11
40. 15
41. 11
42. 16
43. 14
44. 12
45. 12
46. 15
47. 18
48. 13
49. 11
50. 15
51. 11
52. 14
53. 12
54. 17
55. 9
56. 8
57. 10
58. 10
59. 11
60. 12
61. 12
62. 15
63. 16
64. 14
65. 13
66. 11
67. 15
68. 13
69. 13
70. 12
71. 14
72. 14
73. 11
74. 4
75. 9
76. 10
77. 12
78. 10
79. 16
80. 14

Page 13: Counting Patterns

1. **5 10** 15 20 25 30 35 40 45 50 55 **60**
2. **10 20** 30 40 50 60 70 80 90 100 110 **120**
3. **3 5** 7 9 11 13 15 17 19 21 23 **25**
4. **13 19** 25 31 37 43 49 55 61 67 73 **79**
5. **5 9** 13 17 21 25 29 33 37 41 45 **49**
6. **0 3** 6 9 12 15 18 21 24 27 30 **33**
7. **3 10 17** 24 31 38 45 52 **59**
8. **4** + $\boxed{9}$, **13** + $\boxed{9}$, **22** 31 40 49 58 67 **76**

Page 13: Counting Patterns (continued)

9. **9** + [**9**], **18** + [**9**], **27** 36 45 54 63 72 **81**
10. **7 15 23** 31 39 47 55 63 **71**
11. **11 17 23** 29 35 41 47 53 **59**
12. **9 16 23** 30 37 44 51 58 **65**

Page 14: Timed Addition Drill

1. 14
2. 18
3. 20
4. 19
5. 9
6. 10
7. 10
8. 14
9. 10
10. 10
11. 12
12. 15
13. 11
14. 14
15. 16
16. 16
17. 13
18. 17
19. 7
20. 11
21. 9
22. 6
23. 9
24. 9
25. 17
26. 8
27. 15
28. 14
29. 19
30. 11
31. 11
32. 13
33. 11
34. 8
35. 12
36. 14
37. 16
38. 12
39. 18
40. 8
41. 7
42. 6
43. 15
44. 13
45. 12
46. 12
47. 17
48. 12
49. 16
50. 18
51. 5
52. 9
53. 15
54. 7

Page 15: Column Addition

1. 21
2. 20 14 + 6
3. 17 9 + 8
4. 14 9 + 5
5. 20
6. 23
7. 24
8. 26
9. 17
10. 19
11. 21
12. 35 13

Page 16: Two-Digit Addition

A. 95
B. 83
C. 78

1. 39
2. 53
3. 87
4. 99
5. 99
6. 48
7. 48
8. 60
9. 98
10. 89
11. 59
12. 77

Page 16: Two-Digit Addition (continued)

13. 67
14. 77
15. 88
16. 78

Page 17: Three-Digit Addition

1. 979
2. 848
3. 474
4. 505
5. 498
6. 698
7. 531
8. 587
9. 947
10. 899
11. 373
12. 849
13. 935
14. 249
15. 586
16. 575

Page 18: Find the Missing Number

1. 2
2. 3
3. 0
4. 5
5. 3
6. 2
7. 5
8. 0
9. 3
10. 5
11. 4
12. 4
13. 5
14. 4
15. 3
16. 6
17. 6
18. 2
19. 2
20. 4
21. 3
22. 5
23. 0
24. 1
25. 8
26. 7
27. 5
28. 4
29. 4
30. 0
31. 3
32. 3
33. 4
34. 3
35. 6
36. 7
37. 6
38. 3
39. 2
40. 3
41. 3
42. 2
43. 5
44. 8
45. 0
46. 2
47. 1
48. 2
49. 5
50. 1
51. 4
52. 1
53. 6
54. 2
55. 3
56. 3
57. 0
58. 2
59. 0
60. 5
61. 1
62. 2
63. 3
64. 1
65. 3
66. 4
67. 0
68. 0
69. 3
70. 3
71. 8
72. 1
73. 4
74. 3
75. 0
76. 5
77. 6
78. 6
79. 1
80. 3

Page 19: Regroup the Ones

1. 53
2. 95
3. 92
4. 131
5. 123
6. 153
7. 120
8. 117
9. 134
10. 161
11. 110
12. 111
13. 185
14. 160
15. 136
16. 131

Page 20: Regroup and Think Zero

A. 62
B. 105
C. 80
1. 42
2. 100
3. 31
4. 20
5. 51
6. 88
7. 50
8. 84
9. 61
10. 72
11. 31
12. 74
13. 50
14. 74
15. 60
16. 75
17. 41
18. 83
19. 101
20. 23

Page 21: Adding More Numbers

1. 1,478
2. 1,311
3. 1,773
4. 1,290
5. 1,412
6. 1,770

Page 22: Using a Grid

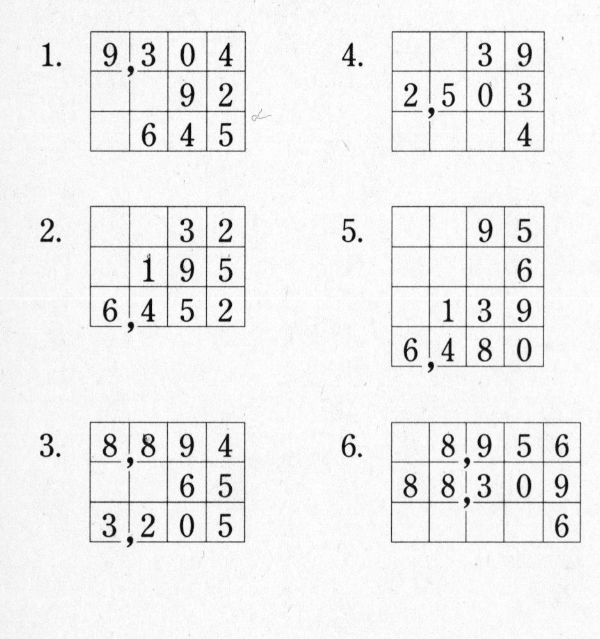

1.

9,	3	0	4
		9	2
	6	4	5

2.

		3	2
	1	9	5
6,	4	5	2

3.

8,	8	9	4
		6	5
3,	2	0	5

4.

		3	9
2,	5	0	3
			4

5.

		9	5
			6
	1	3	9
6,	4	8	0

6.

	8,	9	5	6
8	8,	3	0	9
				6

Page 22: Using a Grid (continued)

7.

2	3,	9	1	4
			5	6
				9
			1	3

8.

		3	9	2
				8
9	4,	1	3	8
			1	4
				4

Page 23: Lining Up Numbers

1. 13,514
2. 7,879
3. 1,110
4. 8,320
5. 7,111
6. 6,290
7. 1,952
8. 9,889

Page 24: Addition Review

1. a) 4,932 b) 609,056 c) 54,962
2. a) 349 b) 9,614 c) 32,100
3. a) 84 b) 146 c) 2,296
4. a) 93 b) 6,840 c) 61,232
5. 4,379,321
6. 75 431 705 9,103 47,147
7. $1{,}932 = (1 \times 1{,}000) + (9 \times 100) + (3 \times 10) + (2 \times 1)$
   $= 1{,}000 + 900 + 30 + 2$
8. 197
9. 8,023
10. 3,212

CHAPTER 3: ADDITION PROBLEM SOLVING

Page 25: Problem-Solving Strategies

1. $9 + $5 = $14
2. $17 + $15 = $32
3. 19 + 8 = 27
4. 13 + 7 = 20
5. 8 + 9 = 17
6. 16 + 9 = 25
7. 6 + 3 = 9
8. $4 + $3 = $7

Page 26: Addition Word Problems

1. 128 + 75 = 203; 203
2. $12,954 + $2,129 = $15,083; $15,083
3. 6,495 + 5,092 = 11,587; 11,587
4. 486 + 294 = 780; 780
5. 24,182 + 6,759 = 30,941; 30,941
6. 48,560 + 25,709 = 74,269; 74,269

CHAPTER 4: SUBTRACTION

Page 27: Meaning of Subtraction

1. a) 5 b) 8 − 3 = 5
2. a) 3 b) 5 − 2 = 3
3. 6 − 3 = 3

Page 28: Practice Helps

1. 1	26. 2	51. 6
2. 2	27. 0	52. 4
3. 5	28. 2	53. 2
4. 4	29. 0	54. 8
5. 0	30. 3	55. 7
6. 2	31. 1	56. 7
7. 1	32. 6	57. 6
8. 2	33. 6	58. 3
9. 2	34. 1	59. 7
10. 5	35. 3	60. 12
11. 8	36. 2	61. 4
12. 1	37. 6	62. 9
13. 5	38. 7	63. 8
14. 3	39. 4	64. 6
15. 5	40. 5	65. 9
16. 4	41. 7	66. 5
17. 3	42. 7	67. 10
18. 4	43. 8	68. 7
19. 3	44. 4	69. 9
20. 2	45. 5	70. 10
21. 0	46. 5	71. 11
22. 8	47. 8	72. 5
23. 1	48. 9	73. 3
24. 0	49. 8	74. 7
25. 3	50. 4	75. 9

Page 28: Practice Helps (continued)

76. 3	78. 10	80. 6
77. 10	79. 5	

Page 29: Timed Subtraction Drill

1. 3	19. 3	37. 7
2. 3	20. 2	38. 11
3. 7	21. 3	39. 7
4. 0	22. 4	40. 4
5. 4	23. 4	41. 11
6. 1	24. 2	42. 0
7. 0	25. 5	43. 9
8. 7	26. 3	44. 5
9. 4	27. 4	45. 5
10. 1	28. 5	46. 8
11. 4	29. 1	47. 4
12. 0	30. 3	48. 2
13. 8	31. 3	49. 8
14. 1	32. 10	50. 10
15. 6	33. 5	51. 1
16. 1	34. 6	52. 6
17. 0	35. 8	53. 8
18. 2	36. 0	54. 6

Page 30: Relating Addition and Subtraction

1. 10 − 4 = 6	11. 2
2. 14 − 5 = 9	12. 9
3. 7 − 4 = 3	13. 15
4. 6 − 1 = 5	14. 6
5. 7 − 7 = 0	15. 8
6. 13 − 8 = 5	16. 24
7. 4 + 7 = 11, so 11 − 4 = 7	17. 14
8. 7 + 3 = 10, so 10 − 3 = 7	18. 97
9. 3 + 9 = 12, so 12 − 9 = 3	19. 29
10. 9 + 0 = 9, so 9 − 0 = 9	20. 8

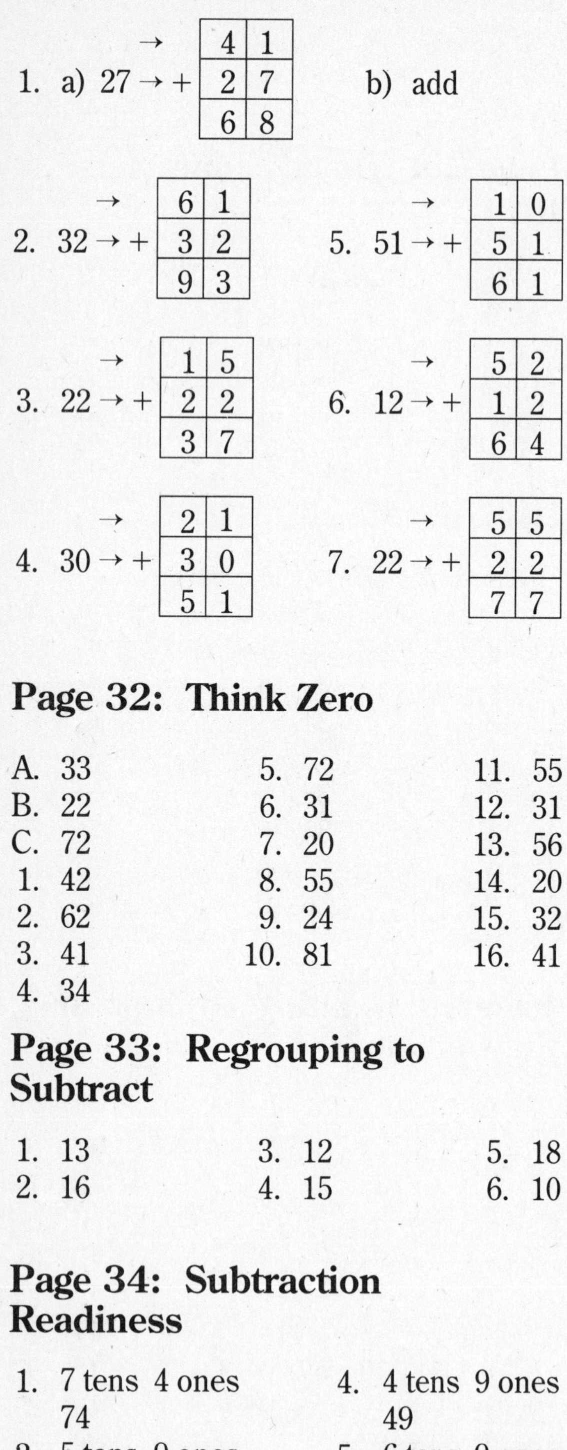

Page 31: Subtracting Tens and Ones

1. a) 27 → +

4	1
2	7
6	8

b) add

2. 32 → +

6	1
3	2
9	3

3. 22 → +

1	5
2	2
3	7

4. 30 → +

2	1
3	0
5	1

5. 51 → +

1	0
5	1
6	1

6. 12 → +

5	2
1	2
6	4

7. 22 → +

5	5
2	2
7	7

Page 32: Think Zero

A. 33
B. 22
C. 72
1. 42
2. 62
3. 41
4. 34
5. 72
6. 31
7. 20
8. 55
9. 24
10. 81
11. 55
12. 31
13. 56
14. 20
15. 32
16. 41

Page 33: Regrouping to Subtract

1. 13
2. 16
3. 12
4. 15
5. 18
6. 10

Page 34: Subtraction Readiness

1. 7 tens 4 ones
   74
2. 5 tens 9 ones
   59
3. 4 tens 8 ones
   48
4. 4 tens 9 ones
   49
5. 6 tens 9 ones
   69
6. 5 tens 6 ones
   56

Page 35: Regrouping Tens to Ones

Step 4. 29
1. 19
2. 18
3. 79
4. 88
5. 57
6. 58
7. 47
8. 58
9. 27
10. 29
11. 45
12. 68

Page 36: Regroup and Subtract

1. 34
2. 19
3. 9
4. 79
5. 36
6. 29
7. 44
8. 6
9. 19
10. 39
11. 27
12. 28

Page 37: Deciding to Regroup

1. Yes; 8 is larger than 7.
2. Yes; 7 is larger than 6.
3. No; 4 is smaller than 9.
4. No; 1 is smaller than 9.
5. Yes; 7 is larger than 5.
6. No; 1 is smaller than 9.

Page 38: Regroup Only When Necessary

1. 14
2. 44
3. 36
4. 44
5. 24
6. 5
7. 39
8. 21
9. 15
10. 40
11. 59
12. 16

Page 39: Subtract and Check

1. 413 → +

1	3	8
4	1	3
5	5	1

2. 229 → +

7	1	3
2	2	9
9	4	2

Page 39: Subtract and Check (continued)

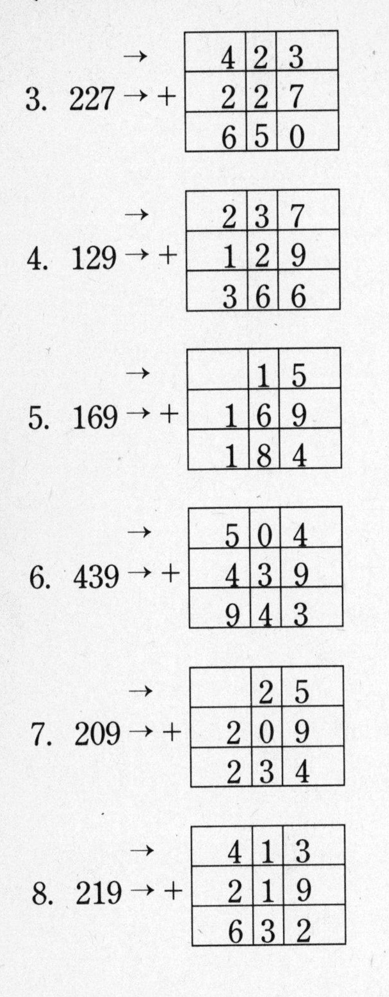

3. 227 → +
4 2 3
2 2 7
6 5 0
4. 129 → +
2 3 7
1 2 9
3 6 6
5. 169 → +
1 5
1 6 9
1 8 4
6. 439 → +
5 0 4
4 3 9
9 4 3
7. 209 → +
2 5
2 0 9
2 3 4
8. 219 → +
4 1 3
2 1 9
6 3 2

Page 40: Regrouping Hundreds and Tens

1. 652
2. 291
3. 173
4. 582
5. 691
6. 245
7. 374
8. 195
9. 702

Page 41: Regrouping More Than Once

1. 383
2. 387
3. 177
4. 189
5. 169
6. 469
7. 93
8. 78
9. 259
10. 279
11. 177
12. 179

Page 42: Regrouping from Zero

A. 335
1. 131
2. 37
3. 184
4. 370
5. 222
6. 416
7. 243
8. 462
9. 204
10. 851
11. 40
12. 501

Page 43: Larger Numbers

A. 4,325
B. 7,736
1. 3,065
2. 1,754
3. 8,538
4. 5,493
5. 5,373
6. 1,154
7. 4,938
8. 4,921
9. 6,409
10. 3,284
11. 2,509
12. 4,948

Page 44: Using a Grid

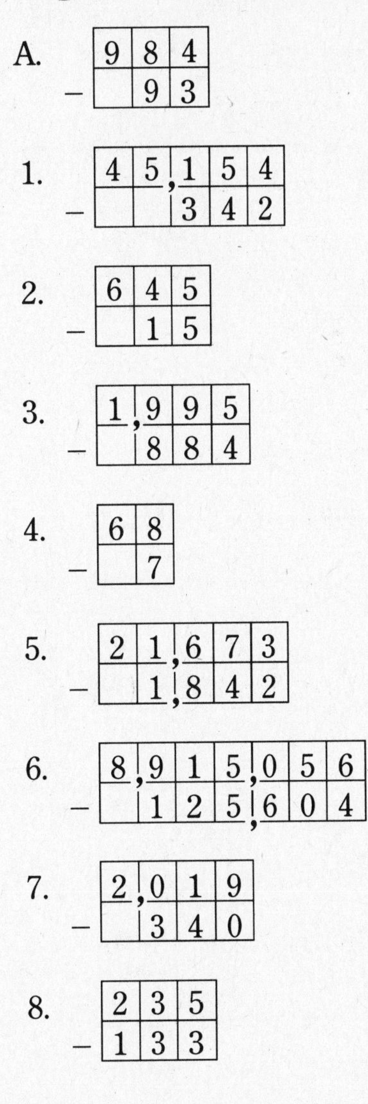

A.
9 8 4
– 9 3
1.
4 5,1 5 4
– 3 4 2
2.
6 4 5
– 1 5
3.
1,9 9 5
– 8 8 4
4.
6 8
– 7
5.
2 1,6 7 3
– 1,8 4 2
6.
8,9 1 5,0 5 6
– 1 2 5,6 0 4
7.
2,0 1 9
– 3 4 0
8.
2 3 5
– 1 3 3

Page 45: Line Up to Subtract

1. 243
2. 462
3. 355
4. 415
5. 4,548
6. 2,350
7. 4,184
8. 3,988
9. 2,660
10. 2,753
11. 51,510
12. 45,044

Page 46: Take Away

A. 61
B. 1,029
1. 66
2. 463
3. 127
4. 289
5. 1,917
6. 647

Page 47: Find the Difference

A. 370
B. 609
1. 225
2. 539
3. 316
4. 177
5. 479
6. 483

Page 48: Subtraction Review

1. If 6 + 7 = 13, then 13 − 7 = 6
2. 83
3. 163
4. 4 tens 5 ones = 3 tens 15 ones
5. Yes; 9 is larger than 8.
6. 32
7. 592
8. 457
9. 379
10. 131
11. 566
12. 368
13. 1,829

Page 49: Putting It All Together

1. <
2. =
3. <
4. <
5. >
6. =
7. =
8. <
9. >
10. =
11. >
12. =
13. =
14. <

CHAPTER 5: SUBTRACTION PROBLEM SOLVING

Page 50: Subtraction Problems

1. 5 − 2 = 3
2. 5 − 2 = 3
3. 6 − 4 = 2
4. 7 − 3 = 4

Page 50: Subtraction Problems (continued)

5. a) 2 b) 2 c) 3 − 1 = 2
6. a) 6 b) 6 c) 10 − 4 = 6

Page 51: Problem-Solving Strategies

1. 7 − 3 = 4
2. 14 − 8 = 6
3. 15 − 7 = 8
4. 25 − 14 = 11
5. 95 − 36 = 59
6. $38 − $10 = $28
7. $125 − $75 = $50
8. $25 − $12 = $13

Page 52: Subtraction Number Sentences

1. 98 − 53 = 45
2. 84 − 36 = 48
3. $125 − $95 = $30
4. 35 − 16 = 19
5. 78 − 15 = 63
6. $194 − $75 = $119
7. $545 − $470 = $75
8. $215 − $175 = $40
9. $75 − $15 = $60
10. $75 − $32 = $43

Page 53: Think About the Facts

Answers may be similar to these.

1. How many students are there in all?
   39 + 16 = 55
2. What is the total of her car and rent payments?
   $260 + $195 = $455
3. How much money does she have after paying bills?
   $537 − $395 = $142
4. How many miles did he drive in all?
   198 + 239 = 437
5. How many empty seats were there?
   497 − 239 = 258
6. How many cards do they have altogether?
   125 + 65 = 190

Page 54: Using Symbols

1. a) $21 - 6 = 15$
   b) $15 + 6 = 21$
   c) $21 > 15$
   d) $15 < 21$
   e) $15 \neq 21$
2. a) $16 - 7 = 9$
   b) $9 + 7 = 16$
   c) $16 > 9$
   d) $9 < 16$
   e) $16 \neq 9$

Page 55: Practice with Symbols

1. a) $3 + 7 = 10$
   b) $10 - 7 = 3$
   c) $10 > 3$
   d) $3 < 10$
   e) $10 \neq 3$
2. a) $30 + 5 = 35$
   b) $35 - 5 = 30$
   c) $35 > 30$
   d) $30 < 35$
   e) $35 \neq 30$

Page 56: Word Problem Review

1. $42 + 57 = 99$
2. \$20 − \$14 = \$6
3. $28 - 17 = 11$
4. \$9 + \$15 = \$24
5. \$125 + \$250 = \$375
6. $250 + 138 = 388$
7. $704 + 435 = 1{,}139$
8. $235 - 68 = 167$
9. $245 - 190 = 55$
10. $136 - 25 = 111$

CHAPTER 6: LIFE-SKILLS MATH

Page 57: Picture Problems

1. $57
2. $133
3. 37 miles
4. 94 miles
5. $144
6. a) 8 hours
   b) 3 hours
   c) 4 hours
   d) 6 hours
   e) 5 hours
7. 26 hours

Page 58: Reading a Map

1. a) 49 miles
   b) 52 miles
   c) 66 miles
   d) 74 miles
   e) 89 miles
   f) 64 miles
   g) 49 miles
2. 202 miles
3. 140 miles
4. 153 miles
5. a) Detroit to Kalamazoo
   b) 13 miles
6. $3,491

Page 59: Using Checks and Calendars

1. Answers will vary.
2. $196
3. One hundred fifty-eight and 00/100
4. $38
5. Sunday
6. 2, 16, and 30
7. June 26
8. 14 days

Page 60: Comparing Prices

1. a) $1,568 b) $1,268
2. $1,265
3. $534
4. $731
5. Answers will vary.

CHAPTER 1: MULTIPLICATION

Page 1: Meaning of Multiplication

1. 5
2. 4
3. 20
4. $5 \times 4 = 20$

Page 2: A Shortcut

1. $5 + 5 + 5 = 15$
   $3 \times 5 = 15$
2. 21, or $3 \times 7 = 21$
3. 16, or $4 \times 4 = 16$
4. 36, or $6 \times 6 = 36$

Page 3: Understanding Multiplication

1. 4
2. 6
3. 24
4. $4 \times 6 = 24$
5. $4 \times 3 = 12$
6. $5 \times 2 = 10$
7. $2 \times 8 = 16$
8. $4 \times 7 = 28$
9. $1 \times 6 = 6$
10. $6 \times 4 = 24$

Page 4: Special Multiplication Rules

1. a) 4×9 is the same as 9×4
   b) 5×7 is the same as 7×5

2. a) $1 \times 4 = 4$
   b) $7 \times 1 = 7$
   c) $1 \times 13 = 13$
   d) $9 \times 1 = 9$

3. a) $0 \times 5 = 0$
   b) $8 \times 0 = 0$
   c) $16 \times 0 = 0$
   d) $0 \times 35 = 0$

Page 5: Memorize Multiplication Facts

1. 4
2. 6
3. 8
4. 10
5. 12
6. 14

Page 5: Memorize Multiplication Facts (continued)

7. 16
8. 18
9. 9
10. 12
11. 15
12. 18
13. 21
14. 24
15. 27
16. 16
17. 20
18. 24
19. 28
20. 32
21. 36
22. 25
23. 30
24. 35
25. 40
26. 45
27. 36
28. 42
29. 48
30. 54
31. 49
32. 56
33. 63
34. 64
35. 72
36. 81

Page 6: Timed Multiplication Drill

1. 9
2. 64
3. 56
4. 35
5. 20
6. 30
7. 8
8. 14
9. 18
10. 25
11. 36
12. 16
13. 36
14. 81
15. 42
16. 7
17. 4
18. 0
19. 16
20. 28
21. 45
22. 48
23. 24
24. 5
25. 63
26. 42
27. 15
28. 12
29. 28
30. 32
31. 27
32. 6
33. 54
34. 21
35. 24
36. 21
37. 8
38. 12
39. 48
40. 18
41. 10
42. 0
43. 1
44. 54
45. 0
46. 56
47. 45
48. 72
49. 32
50. 63
51. 36
52. 72
53. 3
54. 40
55. 8
56. 24

Page 7: Practice Helps

1. 14
2. 45
3. 0
4. 64
5. 40
6. 6
7. 10
8. 24
9. 9
10. 49
11. 30
12. 54
13. 72
14. 18
15. 16
16. 81
17. 42
18. 27

Page 7: Practice Helps (continued)

19. 35
20. 16
21. 36
22. 21
23. 63
24. 36
25. 12
26. 56
27. 4
28. 20
29. 10
30. 9
31. 30
32. 35
33. 48
34. 8
35. 25
36. 54
37. 0
38. 18
39. 27
40. 24
41. 32
42. 28
43. 15
44. 21
45. 12
46. 40
47. 24
48. 12
49. 18
50. 12
51. 8
52. 2
53. 20
54. 3
55. 24
56. 72
57. 0
58. 40
59. 7
60. 8
61. 1
62. 15
63. 48
64. 28
65. 32
66. 18
67. 56
68. 42
69. 16
70. 6
71. 5
72. 0
73. 45
74. 24
75. 0
76. 4
77. 36
78. 15
79. 0
80. 63

Page 8: Find the Missing Numbers

1. 7
2. 3
3. 8
4. 8
5. 1
6. 8
7. 8
8. 5
9. 5
10. 5
11. 3
12. 5
13. 9
14. 7
15. 7
16. 2
17. 4
18. 9
19. 4
20. 0
21. 6
22. 1
23. 9
24. 7
25. 7
26. 3
27. 1
28. 1
29. 4
30. 5
31. 3
32. 8
33. 6
34. 6
35. 6
36. 3
37. 3
38. 2
39. 9
40. 5
41. 5
42. 0
43. 9
44. 9
45. 2
46. 1
47. 9
48. 6
49. 3
50. 2
51. 4
52. 0
53. 5
54. 4
55. 6
56. 2
57. 3

Page 8: Find the Missing Numbers (continued)

58. 5
59. 4
60. 3
61. 7
62. 7
63. 8
64. 9
65. 6
66. 0
67. 2
68. 9
69. 4
70. 1
71. 6
72. 7
73. 5
74. 4
75. 9
76. 6
77. 9
78. 2
79. 4
80. 9

Page 9: Multiplying by One-Digit Numbers

1. 48
2. 90
3. 68
4. 28
5. 86
6. 88
7. 80
8. 48
9. 568
10. 219
11. 248
12. 159
13. 129
14. 255
15. 156
16. 279

Page 10: Multiply and Regroup

1. 120
2. 132
3. 368
4. 329
5. 399
6. 558
7. 588
8. 531
9. 624
10. 315
11. 288
12. 268

Page 11: Hundreds, Tens, and Ones

1. 274
2. 987
3. 624
4. 525
5. 981
6. 896
7. 498
8. 892
9. 898
10. 872
11. 618
12. 678

Page 12: Regrouping Twice

1. 1,172
2. 3,970
3. 3,366
4. 4,476
5. 2,944
6. 3,283
7. 6,606
8. 772
9. 1,812
10. 3,810
11. 5,872
12. 768

Page 13: Multiplying by Two-Digit Numbers

1.
$$\begin{array}{r} 23 \\ \times\ 31 \\ \hline 23 \\ 690 \\ \hline 713 \end{array}$$

2.
$$\begin{array}{r} 12 \\ \times\ 11 \\ \hline 12 \\ 120 \\ \hline 132 \end{array}$$

3.
$$\begin{array}{r} 13 \\ \times\ 23 \\ \hline 39 \\ 260 \\ \hline 299 \end{array}$$

4.
$$\begin{array}{r} 53 \\ \times\ 11 \\ \hline 53 \\ 530 \\ \hline 583 \end{array}$$

5. 169
6. 1,089

7.
$$\begin{array}{r} 43 \\ \times\ 12 \\ \hline 86 \\ 430 \\ \hline 516 \end{array}$$

8. 528
9. 882

Page 14: Multiplying by Ones and Tens

1. 1,260

2.
$$\begin{array}{r} \scriptstyle(3) \\ \scriptstyle\cancel{(2)} \\ 65 \\ \times\ 74 \\ \hline 260 \\ 4550 \\ \hline 4{,}810 \end{array}$$

3. 4,655
4. 2,242
5. 1,664
6. 5,628

Page 15: Multiplying Larger Numbers

1.
$$\begin{array}{r} \scriptstyle(5)\ (6) \\ \scriptstyle\cancel{(4)}\ \cancel{(5)} \\ 368 \\ \times\ 87 \\ \hline 2576 \\ 29440 \\ \hline 32{,}016 \end{array}$$

2. 41,952

3.
$$\begin{array}{r} \scriptstyle(2)\ (1) \\ \scriptstyle\cancel{(6)}\ \cancel{(3)} \\ 574 \\ \times\ 49 \\ \hline 5166 \\ 22960 \\ \hline 28{,}126 \end{array}$$

4. 19,142

5.
$$\begin{array}{r} \scriptstyle(1)\ (1) \\ \scriptstyle\cancel{(1)}\ \cancel{(1)} \\ 823 \\ \times\ 65 \\ \hline 4115 \\ 49380 \\ \hline 53{,}495 \end{array}$$

6. 30,672

Page 16: Multiplying by Multiples of Ten

1. 4,960
2. 1,410
3. 560
4. 5,950
5. 4,440
6. 2,720
7. 8,730
8. 4,400
9. 4,720
10. 31,560
11. 19,720
12. 15,060

Page 17: Larger Number on Top

Step 2.
$$\begin{array}{r} 868 \\ \times\ 92 \\ \hline 1736 \\ 78120 \\ \hline 79{,}856 \end{array}$$

1. 2,310
2. 13,608
3. 64,542
4. 41,736
5. 6,344
6. 17,480

Page 18: Multiplication Review

1. a) 3 b) 6 c) 18 d) $3 \times 6 = 18$
2. 10, or $5 \times 2 = 10$
3. $6 \times 5 = 30$
4. 8
5. 0
6. 25
7. 9
8. 108
9. 270
10. 948
11. 567
12. 231
13. 1,978
14. 25,476
15. 43,500
16. 6,125

CHAPTER 2: MULTIPLICATION PROBLEM SOLVING

Page 19: Writing Number Sentences

1. $7 \times 4 = 28$
2. $\$3 \times 7 = \$21$
3. $\$2 \times 6 = \$12$
4. $15 \times 4 = 60$
5. $5 \times 9 = 45$
6. $7 \times 9 = 63$
7. $\$350 \times 5 = \$1{,}750$
8. $8 \times \$5 = \$40$

Page 20: Writing Labels in Answers

1. $21 \times 7 = 147$ inches
2. $250 \times 3 = 750$ miles
3. $7 \times 3 = 21$ dollars
4. $75 \times 7 = 525$ strands of hair
5. $16 \times 3 = 48$ ounces
6. $9 \times 28 = 252$ pictures

CHAPTER 3: DIVISION

Page 21: Meaning of Division

A. 5
B. 5
1. 12
2. 2
3. 6
4. $12 \div 2 = 6$

Page 22: Division Readiness

1. 20
2. 4
3. 5
4. $20 \div 5 = 4$
5. $36 \div 9 = 4$
6. $30 \div 5 = 6$
7. $18 \div 2 = 9$
8. $36 \div 4 = 9$

Page 23: Relating Multiplication and Division

1. $7 \times 9 = 63$ is the opposite of $63 \div 9 = 7$
2. $5 \times 9 = 45$ is the opposite of $45 \div 5 = 9$
3. $7 \times 8 = 56$ is the opposite of $56 \div 7 = 8$
4. $5 \times 4 = 20$ is the opposite of $20 \div 4 = 5$
5. $3 \times 8 = 24$ is the opposite of $24 \div 8 = 3$
6. $36 \div 6 = 6$ is the opposite of $6 \times 6 = 36$
7. $27 \div 3 = 9$ is the opposite of $3 \times 9 = 27$
8. $49 \div 7 = 7$ is the opposite of $7 \times 7 = 49$
9. $72 \div 8 = 9$ is the opposite of $8 \times 9 = 72$
10. $18 \div 3 = 6$ is the opposite of $6 \times 3 = 18$

Page 24: Division Facts

A. $24 \div 4 = 6$
B. $24 \div 6 = 4$
C. $56 \div 8 = 7$
D. $56 \div 7 = 8$
1. 2
2. 3
3. 4
4. 5
5. 6
6. 7
7. 8
8. 9
9. 3
10. 4
11. 5
12. 6
13. 7
14. 8
15. 9
16. 4
17. 5
18. 6
19. 7
20. 8
21. 9
22. 5
23. 6
24. 7
25. 8
26. 9
27. 6
28. 7
29. 8
30. 9
31. 7
32. 8
33. 9
34. 8
35. 9
36. 9

Page 25: Practice Helps

1. 6
2. 4
3. 9
4. 4
5. 6
6. 7
7. 7
8. 7
9. 5
10. 3
11. 8
12. 6
13. 7
14. 8
15. 9
16. 6
17. 7
18. 6
19. 8
20. 9
21. 9
22. 9
23. 8
24. 5
25. 7
26. 3
27. 9
28. 7
29. 2
30. 3
31. 4
32. 8
33. 4
34. 7
35. 6
36. 5
37. 4
38. 3
39. 0
40. 3
41. 4
42. 5
43. 3
44. 1
45. 5
46. 5
47. 8
48. 5
49. 8
50. 6
51. 4
52. 5
53. 2
54. 3
55. 6
56. 9
57. 2
58. 2
59. 4
60. 2
61. 2
62. 1
63. 2
64. 8
65. 3
66. 1
67. 5
68. 2
69. 9
70. 0
71. 6
72. 1
73. 4
74. 3
75. 11
76. 20
77. 7
78. 6
79. 0
80. 1

Page 26: Two Ways to Show Division

1. $\begin{array}{r} 2 \\ 8\overline{)16} \end{array}$
2. $\begin{array}{r} 5 \\ 3\overline{)15} \end{array}$
3. $\begin{array}{r} 8 \\ 8\overline{)64} \end{array}$
4. $\begin{array}{r} 4 \\ 7\overline{)28} \end{array}$
5. $\begin{array}{r} 5 \\ 9\overline{)45} \end{array}$
6. $\begin{array}{r} 7 \\ 1\overline{)\ 7} \end{array}$
7. $\begin{array}{r} 4 \\ 10\overline{)40} \end{array}$
8. $\begin{array}{r} 7 \\ 8\overline{)56} \end{array}$
9. $\begin{array}{r} 3 \\ 6\overline{)18} \end{array}$
10. $\begin{array}{r} 5 \\ 5\overline{)25} \end{array}$
11. $\begin{array}{r} 1 \\ 9\overline{)\ 9} \end{array}$
12. $\begin{array}{r} 9 \\ 6\overline{)54} \end{array}$
13. $\begin{array}{r} 2 \\ 7\overline{)14} \end{array}$
14. $\begin{array}{r} 6 \\ 8\overline{)48} \end{array}$
15. $\begin{array}{r} 8 \\ 9\overline{)72} \end{array}$
16. $\begin{array}{r} 7 \\ 6\overline{)42} \end{array}$

Page 27: Basic Division

A. 2
B. 1
C. 2 R 1
D. 2 R 1
1. 2 R 3
2. 2 R 1
3. 6 R 1
4. 2 R 4

Think:
8 × 1 = 8
8 × 2 = 16 ✔
8 × 3 = 24

5. 1 R 1
6. 3 R 4
7. 2 R 2
8. 4 R 2
9. 5 R 2
10. 1 R 2
11. 3 R 1
12. 2 R 3

CHAPTER 4: LONG DIVISION

Page 28: Steps for One-Digit Long Division

Step 5:
```
   34
6)204
  18
   24
   24
    0
```

Long division has 5 steps. They are Step 1 **divide,** Step 2 **multiply,** Step 3 **subtract,** Step 4 **compare,** and Step 5 **bring down.**

Page 29: Practice the Steps

```
1.   49
   7)343
     28
      63
      63
       0

2.   17
   9)153
      9
      63
      63
       0

3.   24 R 4
   6)148
     12
      28
      24
       4

4.   37
   4)148
     12
      28
      28
       0

5.   27 R 1
   2)55
     4
     15
     14
      1

6.   91 R 1
   8)729
     72
      09
       8
       1
```

Page 29: Practice the Steps (continued)

```
7.   22 R 2
   4)90
     8
     10
      8
      2

8.   76
   6)456
     42
      36
      36
       0

9.   59
   5)295
     25
      45
      45
       0

10.   31 R 2
    8)250
      24
       10
        8
        2

11.   87 R 1
    7)610
      56
       50
       49
        1

12.   83
    3)249
      24
       09
        9
        0
```

Page 30: Where to Start?

```
1.   61
   7)427
     42
      07
       7
       0

2.   434 R 1
   2)869
     8
     06
      6
      09
       8
       1

3.   192
   4)768
     4
     36
     36
      08
       8
       0

4.   72
   9)648
     63
      18
      18
       0

5.   211
   4)844
     8
     04
      4
      04
       4
       0

6.   131
   4)524
     4
     12
     12
      04
       4
       0
```

Page 31: Zeros in the Answer

1. $4\overline{)816}$ = 204
   8
   016
   16
   0

2. $3\overline{)270}$ = 90
   27
   00
   0
   0

3. $7\overline{)753}$ = 107 R 4
   7
   053
   49
   4

4. 80 R 6
5. 20
6. 109 R 4
7. 180 R 4
8. 109
9. 407 R 1

Page 32: Using a Grid

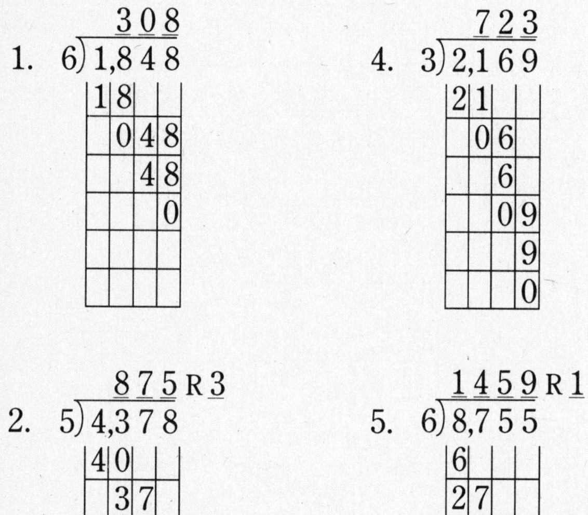

1. $6\overline{)1{,}848}$ = 308
   18
   048
   48
   0

4. $3\overline{)2{,}169}$ = 723
   21
   06
   6
   09
   9
   0

2. $5\overline{)4{,}378}$ = 875 R 3
   40
   37
   35
   28
   25
   3

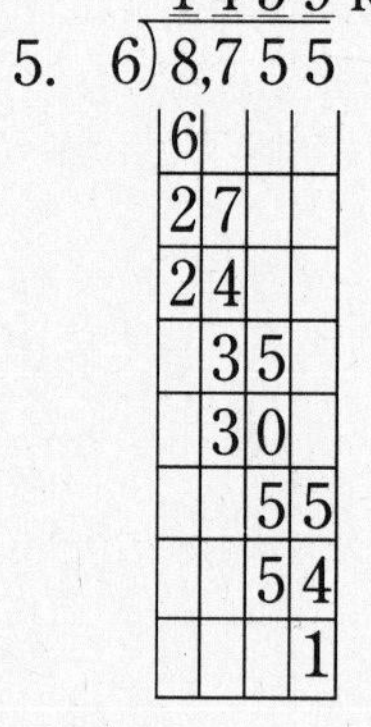

5. $6\overline{)8{,}755}$ = 1459 R 1
   6
   27
   24
   35
   30
   55
   54
   1

3. $9\overline{)5{,}283}$ = 587
   45
   78
   72
   63
   63
   0

Page 32: Using a Grid (continued)

6. $7\overline{)5{,}621}$ = 803
   56
   02
   0
   21
   21
   0

8. $8\overline{)8{,}347}$ = 1043 R 3
   8
   034
   32
   27
   24
   3

7. $5\overline{)9{,}845}$ = 1969
   5
   48
   45
   34
   30
   45
   45
   0

9. $9\overline{)8{,}055}$ = 895
   72
   85
   81
   45
   45
   0

Page 33: Dividing by One Digit

1. 1,200
2. 1,681
3. 1,151
4. 305
5. 254
6. 304
7. 479
8. 406
9. 190
10. 1,253
11. 909
12. 450

Page 34: Steps for Two-Digit Division

A. $90\overline{)2468}$ = 27 R 38
   180
   668
   630
   38

Page 35: Dividing by Multiples of Ten

1. $40\overline{)346}$ = 8 R 26
   320
   26

2. $20\overline{)196}$ = 9 R 16
   180
   16

Page 35: Dividing by Multiples of Ten (continued)

3. $\begin{array}{r} 5 \text{ R } 8 \\ 10\overline{)58} \\ \underline{50} \\ 8 \end{array}$

4. $\begin{array}{r} 2 \text{ R } 68 \\ 90\overline{)248} \\ \underline{180} \\ 68 \end{array}$

Think:
90 × 2 = 180
90 × 3 = 270

5. $\begin{array}{r} 7 \text{ R } 9 \\ 50\overline{)359} \\ \underline{350} \\ 9 \end{array}$

Think:
50 × 6 = 300
50 × 7 = 350

6. $\begin{array}{r} 7 \text{ R } 10 \\ 70\overline{)500} \\ \underline{490} \\ 10 \end{array}$

7. $\begin{array}{r} 2 \text{ R } 70 \\ 80\overline{)230} \\ \underline{160} \\ 70 \end{array}$

8. $\begin{array}{r} 1 \text{ R } 44 \\ 50\overline{)94} \\ \underline{50} \\ 44 \end{array}$

9. $\begin{array}{r} 3 \text{ R } 23 \\ 40\overline{)143} \\ \underline{120} \\ 23 \end{array}$

10. $\begin{array}{r} 3 \text{ R } 9 \\ 30\overline{)99} \\ \underline{90} \\ 9 \end{array}$

Page 36: Estimating Is Key

1. 70
2. 90
3. 50
4. 40
5. 20
6. 20
7. Estimate: 90 into 207 → 2
8. Estimate: 90 into 330 → 3
9. Estimate: 70 into 496 → 7
10. Estimate: 60 into 363 → 6

Page 37: Estimate the Quotient

1. $\begin{array}{rr} & 1 \\ & 31\overline{)52} \\ 30 & \underline{31} \\ & 21 \end{array}$ ☑ < 52 ☑ < 31

2. $\begin{array}{rr} & 2 \\ & 32\overline{)76} \\ 30 & \underline{64} \\ & 12 \end{array}$ ☑ < 76 ☑ < 32

3. $\begin{array}{rr} & 6 \\ & 29\overline{)190} \\ 30 & \underline{174} \\ & 16 \end{array}$ ☑ < 190 ☑ < 29

Page 37: Estimate the Quotient (continued)

4. $\begin{array}{rr} & 6 \\ & 81\overline{)526} \\ 80 & \underline{486} \\ & 40 \end{array}$ ☑ < 526 ☑ < 81

5. $\begin{array}{rr} & 9 \\ & 58\overline{)562} \\ 60 & \underline{522} \\ & 40 \end{array}$ ☑ < 562 ☑ < 58

6. $\begin{array}{rr} & 6 \\ & 48\overline{)312} \\ 50 & \underline{300} \\ & 12 \end{array}$ ☑ < 312 ☑ < 48

7. $\begin{array}{rr} & 3 \\ & 18\overline{)63} \\ 20 & \underline{54} \\ & 9 \end{array}$ ☑ < 63 ☑ < 18

8. $\begin{array}{rr} & 4 \\ & 79\overline{)386} \\ 80 & \underline{316} \\ & 70 \end{array}$ ☑ < 386 ☑ < 79

9. $\begin{array}{rr} & 3 \\ & 72\overline{)266} \\ 70 & \underline{216} \\ & 50 \end{array}$ ☑ < 266 ☑ < 72

Page 38: Adjust When Necessary

1. $\begin{array}{r} 7 \\ \not{6} \\ 38\overline{)268} \\ \underline{266} \\ 2 \end{array}$ ☑ ☑

2. $\begin{array}{r} 4 \\ \not{5} \\ 33\overline{)159} \\ \underline{132} \\ 27 \end{array}$ ☑ ☑

3. $\begin{array}{r} 7 \\ 29\overline{)209} \\ \underline{203} \\ 6 \end{array}$ ☑ ☑

4. $\begin{array}{r} 8 \\ \not{9} \\ 89\overline{)713} \\ \underline{712} \\ 1 \end{array}$ ☑ ☑

Page 38: Adjust When Necessary (continued)

5. 67)275 — quotient 4
   268 ☑
   7 ☑

6. 33)159 — quotient ~~5~~ 4
   132 ☑
   27 ☑

7. 27)172 — quotient 6
   162 ☑
   10 ☑

8. 33)198 — quotient ~~5~~ 6
   198 ☑
   0 ☑

9. 92)366 — quotient ~~4~~ 3
   276 ☑
   90 ☑

Page 39: Decide Where to Start

1. 41)307 — X
2. 36)621 — XX
3. 45)443 — X
4. 62)621 — XX
5. 14)286 — XX
6. 4)832 — XXX
7. 16)174 — XX
8. 91)103 — X
9. 7)708 — XXX
10. 83)868 — XX

Page 40: Round to Divide

1. 18)49 — quotient 2
   [20] 36 ☑
   13 ☑

2. 47)230 — quotient 4
   [50] 188 ☑
   42 ☑

3. 53)263 — quotient 4
   [50] 212 ☑
   51 ☑

4. 88)326 — quotient 3
   [90] 264 ☑
   62 ☑

Page 40: Round to Divide (continued)

5. 54)395 — quotient 7
   [50] 378 ☑
   17 ☑

6. 26)81 — quotient 3
   [30] 78 ☑
   3 ☑

7. 19)168 — quotient 8
   [20] 152 ☑
   16 ☑

8. 34)128 — quotient 3
   [30] 102 ☑
   26 ☑

9. 59)472 — quotient 8
   [60] 472 ☑
   0 ☑

10. 67)297 — quotient 4
    [70] 268 ☑
    29 ☑

11. 73)432 — quotient 5
    [70] 365 ☑
    67 ☑

12. 23)198 — quotient 8
    [20] 184 ☑
    14 ☑

Page 41: Digits in the Quotient

1. 63)9,253 — XXX
2. 9)68,259 — X XXX
3. 189)2,237 — XX
4. 98)1,000 — XX
5. 602)16,856 — XX
6. 213)943 — X
7. 90)800 — X
8. 743)1,649 — X
9. 5)4,096 — XXX
10. 73)8,506 — XXX
11. 8)596 — XX
12. 43)2,476 — XX
13. 49)39,741 — XXX
14. 19)37,520 — X XXX
15. 973)6,190 — X

Page 42: Estimate the Answer

1. 7
2. 2,000
3. 30
4. 700
5. 9
6. 1,000
7. 300
8. 40

Page 43: Divide by Two Digits

1.
```
        548
  72) 39,456
      360
       345
       288
        576
        576
          0
```
2. 839
3. 16 R 7
4. 782
5. 79 R 1
6. 492

Page 44: Practice Helps

1.
```
       107
  23) 2,461
      2 3
        16
         0
        161
        161
          0
```
2. 306
3. 408
4. 19
5. 217
6. 6 R 19
7. 108 R 9
8. 408
9. 903

Page 45: Division Review

1. 36 ÷ 6 = 6
2. 7 × 8 = 56 is the opposite of 56 ÷ 8 = 7
3.
```
      7
  9) 63
```
4. 3 R 2
5. 28
6. 21
7. 207
8. 701
9. 8 R 16
10. 50
11. 8 R 4 round to } 80
12. 80 R 20 round to } 20
13. 40

Page 46: Putting It All Together

1. 120 < 126
2. 528 < 588
3. 92 = 92
4. 8 > 7

Page 46: Putting It All Together (continued)

5. 2,304 = 2,304
6. 473 > 453
7. 78 < 81
8. 4,224 > 4,053
9. 99 < 100
10. 658 < 703
11. 72 = 72
12. 38 > 36

CHAPTER 5: DIVISION PROBLEM SOLVING

Page 47: Writing Number Sentences

1. \$45 ÷ 3 = \$15
2. 24 ÷ 3 = 8 hours
3. 8 ÷ 4 = 2 pieces
4. \$12 ÷ 3 = \$4
5. \$36 ÷ 9 = \$4
6. \$24 ÷ 2 = \$12

Page 48: Writing Labels in Answers

1. \$175 ÷ \$25 = 7 weeks
2. \$57 ÷ 3 = 19 dollars
3. 36 ÷ 6 = 6 sacks
4. 440 ÷ 4 = 110 yards
5. 18 ÷ 2 = 9 calls; calls
6. \$12 ÷ 4 = 3 dollars; dollars
7. \$64 ÷ \$16 = 4 shirts; shirts
8. \$15 ÷ 3 = 5 dollars; dollars

Page 49: Think About the Remainder

1. 2 R 4; 3 tables
2. 31 R 1; 31 boards
3. 2 R 1; 2 cakes
4. 3 R 1; 4 Popsicles
5. 6 R 10; 6 cars
6. 12 R 3; 13 nickels

CHAPTER 6: MIXED PROBLEM SOLVING

Page 50: Mixed Multiplication and Division

1. 7 × 16 = 112 boxes

Page 50: Mixed Multiplication and Division (continued)

2. $24 \div 8 = 3$ feet
3. $75 \times 5 = 375$ miles
4. 4¢ × 15 = 60¢
5. $135 \div 15 = 9$ weeks
6. \$65 × 3 = \$195
7. $75 \times 5 = 375$ times
8. $285 \div 15 = 19$ payments
9. \$25 × 15 = \$375
10. $270 \div 5 = 54$ miles

Page 51: Choose the Operation

1. division
2. subtraction
3. multiplication
4. addition
5. ×
6. ×
7. +
8. −
9. ÷
10. +

Page 52: Write the Question

Answers will vary.

Page 53: Decide the Operation

1. \$25 + \$17 = \$42
2. \$10 − \$3 = \$7
3. $115 \div 23 = 5$ buses
4. $7 \times 5 = 35$ minutes
5. $485 - 138 = 347$ miles
6. $28 \div 7 = 4$ students
7. $500 \div 25 = 20$ gallons
8. \$12 × 5 = \$60

Page 54: Mixed Practice

1. 485 + 1,340 = \$1,825; 1,825
2. 18 × 9 = 162; 162
3. 9 × \$20 = \$180; 180
4. \$215 − \$119 = \$96; 96
5. \$295 + \$15 = \$310; 310
6. \$279 ÷ 3 = \$93; 93
7. 15 × 8 = 120; 120
8. 12 × 4 = 48; 48

Page 55: Looking at Word Problems

1. a) 175
   b) 175
   c) 175
   d) 175
   e) 175
2. a) 90
   b) 90
   c) 90
3. a) 5
   b) 5
   c) 5
   d) 5
   e) 5

Page 56: Apply Your Skills

1. D
2. B
3. A
4. C
5. 15
6. 15
7. $150 > 135$
8. $135 < 150$
9. $135 \neq 150$
10. \$135 − \$78 = \$57; \$57
11. \$15 ÷ 3 = \$5; \$5

CHAPTER 7: LIFE-SKILLS MATH

Page 57: Everyday Math

1. a) 6
   b) 12

2. \$140

3. a) **2** × 15 = **\$30**
   b) 3 × **\$8** = **\$24**
   c) \$30 + \$24 = \$54

4. a) \$13
   b) 3 × **\$13** = **\$39**

5. \$375

6. a) 4
   b) **4** × \$5 = **\$20**

Page 58: Picture Problems

1. 96 color prints
2. \$96
3. \$28
4. 384 miles per hour
5. \$6
6. 375 miles

Page 59: Weekly Pay

	Hours Worked	Paid
1.	7	$63
2.	5	$45
3.	4	$36
4.	8	$72
5.	6	$54
6.	30	$270

7. 30 hours
8. $270
9. a) $18
   b) $18
   c) $27
10. 2 hours
11. $1,080
12. 26 weeks

Page 60: Weekly Sales

	Unit Price	Amount
1.	$260	$2,080
2.	$578	$2,890
3.	$53	$1,378
4.		$6,348

5. 13 months
6. 17 months
7. 51 hours
8. 6,348; 6,348 ÷ 2 = $3,174

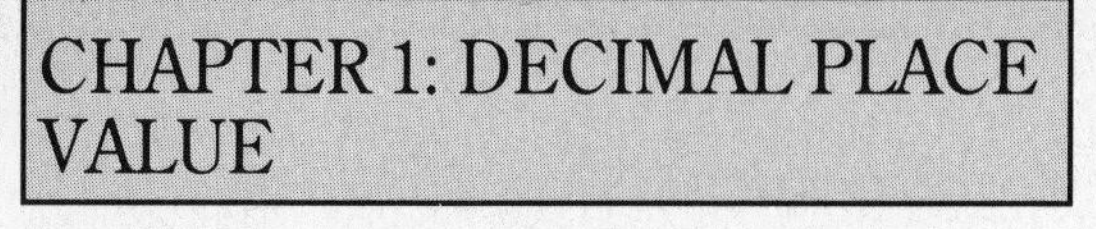

CHAPTER 1: DECIMAL PLACE VALUE

Page 1: Meaning of Tenths

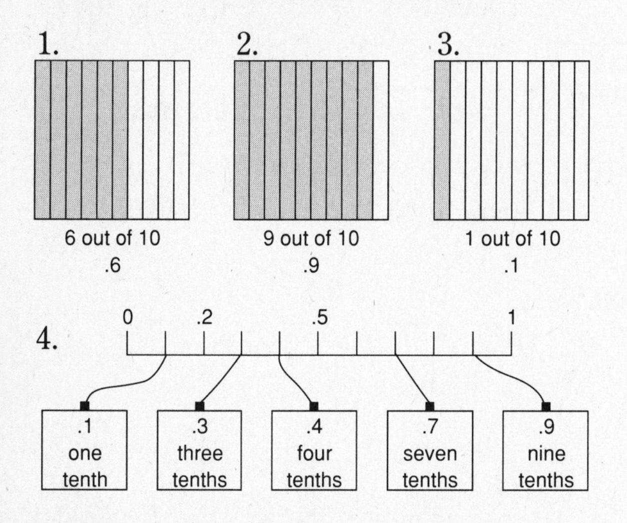

Page 2: Comparing Tenths

1. greater
2. less
3. less
4. less
5. greater
6. greater
7. greater
8. greater
9. smaller
10. larger

Page 3: Meaning of Hundredths

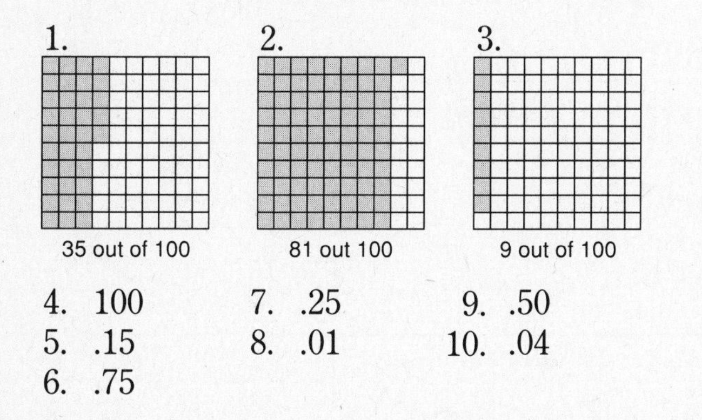

4. 100
5. .15
6. .75
7. .25
8. .01
9. .50
10. .04

Page 4: Comparing Hundredths

1. greater
2. greater
3. greater
4. less

Page 4: Comparing Hundredths (continued)

5. less
6. greater
7. less
8. less

Page 5: Meaning of Thousandths

1. .035
2. .069
3. .350
4. .936
5. .525
6. 250 out of 1,000 .250
7. 40 out of 1,000 .040
8. 485 out of 1,000 .485
9. 806 out of 1,000 .806
10. 15 out of 1,000 .015
11. 685 out of 1,000 .685
12. 167 out of 1,000 .167

Page 6: Comparing Thousandths

1. less
2. greater
3. greater
4. less
5. greater
6. greater
7. less
8. greater
9. greater
10. less

Page 7: Place-Value Readiness

hundreds	tens	ones		tenths	hundredths	thousandths
8	7	0	.	2	1	4

A. 8 7 0 . 2 1 4

1. ones
2. hundredths
3. tenths
4. thousandths
5. tens
6. 0
7. 8
8. 7
9. 1
10. 2

Page 8: Place Values

1. a) 7 tens
   1 one
   6 tenths
2. a) 9 tens
   b) 4 ones
   c) 3 tenths
   d) 5 hundredths

Page 8: Place Values (continued)

3. a) 8 tens
   b) 9 ones
   c) 4 tenths
   d) 6 hundredths
   e) 3 thousandths

Page 9: Reading Decimals

			and				
1. 71.647	7	1	.	6	4	7	71 and 647 thousandths
2. 50.94	5	0	.	9	4		50 and 94 hundredths
3. 1.6		1	.	6			1 and 6 tenths
4. 16.345	1	6	.	3	4	5	16 and 345 thousandths
5. 3.7		3	.	7			3 and 7 tenths
6. 8.04		8	.	0	4		8 and 4 hundredths
7. 17.39	1	7	.	3	9		17 and 39 hundredths
8. 98.507	9	8	.	5	0	7	98 and 507 thousandths
9. 7.53		7	.	5	3		7 and 53 hundredths
10. 84.2	8	4	.	2			84 and 2 tenths
11. 4.008		4	.	0	0	8	4 and 8 thousandths
12. 7.45		7	.	4	5		7 and 45 hundredths
13. 6.19		6	.	1	9		6 and 19 hundredths
14. 29.133	2	9	.	1	3	3	29 and 133 thousandths
15. 2.4		2	.	4			2 and 4 tenths

Page 10: Write the Place Value

1. ones
2. tenths
3. hundredths
4. tenths
5. ones
6. hundredths
7. tens
8. tenths
9. hundreds
10. tenths
11. hundredths
12. tens
13. thousandths
14. hundreds
15. tenths
16. thousandths
17. hundredths
18. tens

Page 11: Identify the Digit

1. 927.452
2. 469.36
3. 2,314.067
4. .94
5. 97.8
6. .056
7. 15.04
8. .458
9. 321.45
10. 532.497

Page 12: Zeros in Decimals

1. .230, .2300
2. .0460, .046
3. .10, .100
4. 3.50, 3.500
5. .070, .07
6. 9.080, 9.08

Page 13: Writing Zeros to Hold Place Values

1. .037
2. .06
3. .9
4. .16
5. .427
6. .001
7. 1.014
8. 16.09
9. 427.3
10. 7.007
11. .05
12. .3

Page 14: Words to Decimals

				and			
1. Six and four tenths			6	.	4		
2. Fifteen and six hundredths		1	5	.	0	6	
3. Three hundred forty-five thousandths				.	3	4	5
4. One hundred and twenty-one hundredths	1	0	0	.	2	1	
5. Eight hundred ninety-four and six tenths	8	9	4	.	6		
6. Seven thousandths				.	0	0	7
7. Ninety-three and fifty-four thousandths		9	3	.	0	5	4
8. Seven and four tenths			7	.	4		
9. Two hundred seven and seventy-one hundredths	2	0	7	.	7	1	
10. Three thousandths				.	0	0	3
11. Five hundred thirty-nine thousandths				.	5	3	9
12. Nine hundred forty-five and three hundredths	9	4	5	.	0	3	
13. Three and eight thousandths			3	.	0	0	8
14. Eight tenths				.	8		
15. One hundred five and one hundredth	1	0	5	.	0	1	

Page 15: Find the Number

1. d
2. b
3. a
4. c
5. c
6. b
7. a
8. d
9. c
10. d
11. a
12. b
13. b
14. c
15. d
16. a
17. b
18. c
19. d
20. a

Page 16: Expanding Decimals

1. .73 = .7 + .03
2. .984 = .9 + .08 + .004
3. 8.275 = 8 + .2 + .07 + .005
4. .761 = .7 + .06 + .001
5. 75.94 = 70 + 5 + .9 + .04
6. 5.693 = 5 + .6 + .09 + .003
7. .94 = .9 + .04
8. 4.386 = 4 + .3 + .08 + .006

Page 17: Comparing Decimals

1. >
2. <
3. =
4. >
5. <
6. <

Page 18: Ordering Decimals

1.	4.2	4.25	4.279
2.	4.9	5.03	5.841
3.	7.01	7.032	7.4
4.	15.98	15.981	17.8
5.	94.3	94.301	94.31
6.	13.05	13.5	13.58
7.	3.05	3.50	5.30
8.	.378	3.49	3.5

CHAPTER 2: ADDITION

Page 19: Steps for Addition

A. 6.1
B. 12.2
1. 7.7
2. 10.5
3. 5.7
4. 13.2
5. 15.1
6. 8.3
7. 11.3
8. 12.3
9. 11.0

Page 20: Adding Tenths, Hundredths, and Thousandths

A.
(1)
0.74
+ 0.32
1.06

B.
(1)
0.845
+ 0.653
1.498

1.
(1)
0.5
+ 0.6
1.1

2. 1.0
3. 1.2
4. 1.61
5. .87
6. .11
7. 1.062
8. .502
9. 1.626

Page 21: Adding Mixed Decimals

A.
(1) (1)
9.67
+ 5.94
15.61

B.
7.03
+ 9.25
16.28

1.
(1) (1)
3.97
+ 8.37
12.34

2. 10.09
3. 12.00
4. 11.61
5. .8.55
6. 13.61
7. 6.37
8. 6.71
9. 16.53

Page 22: Think Zero

A.
(1) (1)
47.8
+ 2.3
50.1

B.
(1) (1)
397.2
+ 85.4
482.6

1.
(1) (1)
97.8
+ 3.2
101.0

2. 17.1
3. 47.5
4. 67.0
5. 39.9
6. 49.9
7. 905.7
8. 543.6
9. 41.7

Page 23: Zeros as Placeholders

A.
(1)
5.48
+ 9.80
15.28

B.
(1) (1)
34.60
+ 9.47
44.07

C.
(1)
4.932
+ 1.400
6.332

1.
(1)
7.45
+ 2.90
10.35

2. 10.33
3. 44.16
4. 33.05
5. 71.88
6. 6.76
7. 13.074
8. 40.419
9. 10.583

Page 24: Writing Decimal Points

A. 61.
B. 522.
C. 144.
D. 2.
1. 11.38
2. 24.93
3. 33.9
4. 9.16
5. 22.75
6. 16.63

Page 25: Write Zeros When Necessary

A. 80.8
B. 386.32
1. 9.9
2. 19.99
3. 27.401
4. 17.94
5. 7.56
6. 32.93
7. 14.91
8. 19.7
9. 16.10

Page 26: Lining Up Decimals

```
1.   613.049
        .620
        .953
   +    .900

2.     2.950
      62.000
      45.390
   +   3.846

3.   312.749
      51.090
   +   2.600

4.    $21.40
       $5.07
   +  $51.00

5.     5.460
      93.700
       2.467
   +  45.060

6.     3.200
        .985
        .460
   +   9.000
```

Page 27: Practice Helps

1. 28.543
2. 18.87
3. 15.526
4. 6.60
5. 6.86
6. 82.53
7. 9.849
8. 121.87

Page 28: Addition Review

1. 2.040
2. 18.032
3. <
4. 7.05, 7.105, 7.15
5. .549
6. 1.3
7. 6.228
8. 2.77
9. 62.357
10. 10.334
11. 3.231
12. 17.046

CHAPTER 3: ADDITION PROBLEM SOLVING

Page 29: Decimals and Money

1. c
2. b
3. f
4. d
5. a
6. e

Page 30: Working with Money

1. $5.00 + .25 + .05 + .01 + .01 = $5.32
2. $.05 + .10 + .10 + .05 + .25 = $.55
3. $.25 + .25 + .25 + .10 + .05 = $.90
4. $10.00 + .10 + .10 + .10 + .01 = $10.31

Page 31: Writing Number Sentences

1. $7.45 + $3.15 = $10.60; 10.60
2. $13.05 + $5.62 = $18.67; 18.67
3. $75.11 + $9.28 = $84.39; 84.39
4. $45.00 + $7.99 = $52.99; 52.99
5. $.68 + $6.45 = $7.13; 7.13
6. $9.98 + $15.00 = $24.98; 24.98

Page 32: Number Sentences

1. $17.35 + $6.92 = $24.27
2. $7.92 + $.32 = $8.24
3. .45 + .39 = .84 pound
4. $25.50 + $19.75 = $45.25
5. 3.4 + .6 = 4.0 miles
6. $14.35 + $9.35 = $23.70

CHAPTER 4: SUBTRACTION

Page 33: Steps for Subtraction

A. 4.2
B. .34
1. .2
2. .4
3. .6
4. 3.3
5. 1.3
6. 5.7
7. .11
8. .23
9. .21

Page 34: Regrouping

A. .07
B. 1.8
C. 3.92
1. .08
2. .38
3. .23
4. 5.8
5. 4.8
6. 1.7
7. 7.75
8. 1.56
9. 4.82

Page 35: Mastering Regrouping

A. .609
B. .578
C. 3.36
1. .288
2. .358
3. .029
4. .399
5. .157
6. .267
7. 5.49
8. 3.17
9. 1.39

Page 36: More Regrouping

A. 3.39
B. 5.28
C. 1.97
1. 4.34
2. 8.57
3. 6.69
4. 7.15
5. 5.02
6. 3.72
7. 1.54
8. 2.67
9. 6.79

Page 37: Zeros Help

1. .484
2. .463
3. 4.345
4. .062
5. 7.093
6. .261
7. .296
8. 7.062
9. .417
10. .195
11. .492
12. 4.637

Page 38: Zeros at Work

A. 85.28
B. 64.27
C. 34.47
1. 46.86
2. 72.49
3. 40.66
4. 46.98
5. 51.38
6. 22.75
7. 50.15
8. 2.27
9. 63.65

Page 39: Subtracting Decimals from Whole Numbers

A. 6.5
B. 1.18
C. 6.937
1. 1.9
2. 6.5
3. 14.3
4. 2.57
5. 17.94
6. 14.18
7. 2.982
8. 8.287
9. 4.976

Page 40: Subtracting Whole Numbers from Mixed Decimals

A. 25.5
B. 29.43
C. 52.849

Page 40: Subtracting Whole Numbers from Mixed Decimals (continued)

1. 9.8
2. 46.1
3. 37.3
4. 37.19
5. 165.70
6. 8.62
7. 7.691
8. 18.512
9. 57.822

Page 41: Watch Out for Whole Numbers

A. 5.86
B. 67.2
C. $\overset{2\,14\,15}{\cancel{3}\cancel{5}.\cancel{5}} - 8.9 = 26.6$
1. 7.24
2. 63.8
3. 4.22
4. 24.65
5. 13.7
6. 69.5
7. 9.70
8. 93.4
9. 6.08

Page 42: Add or Subtract?

1. 9.4
2. 51.9
3. 271.43
4. 38.86
5. 8.53
6. 11.74
7. 46.13
8. 20.94
9. 27.17
10. 54.9
11. 1.01
12. 710.67

Page 43: Putting It All Together

1. $14.5 < 14.6$
2. $7.82 > 7.6$
3. $32.73 < 33.2$
4. $67.4 > 66.73$
5. $47.9 > 47.605$
6. $30.4 = 30.4$
7. $142.51 < 144.88$
8. $1.29 = 1.29$

Page 44: Subtraction Review

1. .24
2. 2.4
3. .08
4. 1.19
5. .059
6. .41
7. 1.15
8. .558
9. .53
10. 26.275
11. 29.34
12. .633

CHAPTER 5: SUBTRACTION PROBLEM SOLVING

Page 45: Number Sentences

1. 2.4 − .21 = 2.19 ounces
2. 118.6 + 69.8 = 188.4 miles

Page 45: Number Sentences (continued)

3. $32.45 – $9.50 = $22.95
4. $38.90 – $4.50 = $34.40
5. 15.8 – 2.9 = 12.9 gallons
6. 1.53 + .75 = 2.28 pounds
7. $75.95 – $22.75 = $53.20
8. $125.00 – $95.00 = $30.00

Page 46: Does the Answer Make Sense?

1. 101.4° – 98.6° = 2.8°; 2.8
2. 172.8 + 39.6 = 212.4; 212.4
3. $99.95 – $25.00 = $74.95; 74.95
4. $20.00 – $14.35 = $5.65; 5.65
5. $258.15 – $217.92 = $40.23; 40.23
6. 31.4 + 43 = 74.4; 74.4

Page 47: Using Symbols

1. a) $9.40 + $5.49 = $14.89
   b) $14.89 – $5.49 = $9.40
   c) $14.89 > $9.40
   d) $9.40 < $14.89
   e) $14.89 ≠ $9.40
2. a) $17.19 + $28.14 = $45.33
   b) $45.33 – $28.14 = $17.19
   c) $45.33 > $17.19
   d) $17.19 < $45.33
   e) $45.33 ≠ $17.19

Page 48: Write a Question

Questions should be similar to these:

1. a) How much did Leslie and Dawn spend altogether?
   b) How much more did Leslie spend than Dawn?

2. a) Sean saved how much more than his sister?
   b) How much did Sean and his sister save in all?

Page 48: Write a Question (continued)

3. a) What was the total cost of the groceries and gasoline?
   b) The groceries cost how much more than the gasoline?

4. a) The jacket cost how much more than the sweater?
   b) What was the combined cost of the jacket and sweater?

5. a) Mr. Alatalo and Mr. Forbes saved how much altogether?
   b) Mr. Alatalo saved how much more than Mr. Forbes?

6. a) Mr. Fenwick paid how much in taxes both years combined?
   b) How much more did Mr. Fenwick pay in taxes this year than last?

Page 49: Decide to Add or Subtract

Answers will vary.

Page 50: Mixed Addition and Subtraction

1. $65.34 + $14.95 = $80.29; 80.29
2. $5.00 – $2.91 = $2.09; 2.09
3. 256.3 + 136.9 = 393.2; 393.2
4. 5.4 + 3.2 = 8.6; 8.6
5. $45.32 – $36.50 = $8.82; 8.82
6. $5.00 – $3.25 = $1.75; 1.75
7. $29.15 + $8.11 = $37.26; 37.26
8. $10.00 – $2.25 = $7.75; 7.75

Page 51: Two-Step Word Problems

A. $4.14
B. $15.86

1. a) $66.74 [+]
   b) $8.26 [–]

Page 51: Two-Step Word Problems (continued)

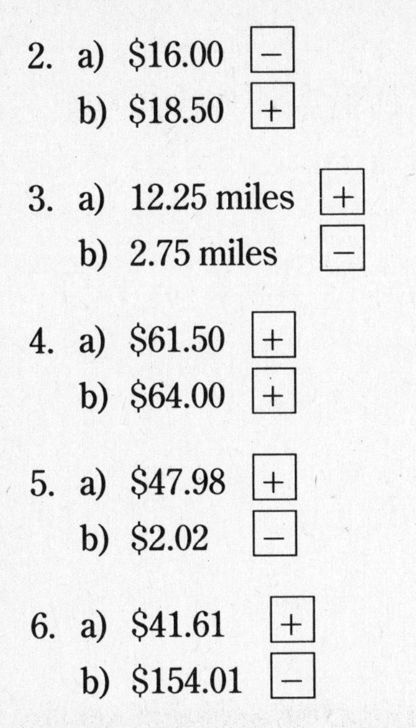

2. a) $16.00 [−]
   b) $18.50 [+]

3. a) 12.25 miles [+]
   b) 2.75 miles [−]

4. a) $61.50 [+]
   b) $64.00 [+]

5. a) $47.98 [+]
   b) $2.02 [−]

6. a) $41.61 [+]
   b) $154.01 [−]

Page 52: Two-Step Problem Solving

Sample Problems

1. a) How much did Danielle loan her friend and spend altogether?
   $8.50 + $12.95 = $21.45

   b) How much did she have left?
   $25.00 − $21.45 = $3.55

2. a) How much were the two rings?
   $47.99 + $47.99 = $95.98

   b) How much did she spend altogether?
   $95.98 + $15.95 = $111.93

3. a) The three deposits total how much?
   $84.50 + $196.38 + $36.47 = $317.35

   b) How much does Lisa have in her savings account?
   $317.35 + $436.65 = $754.00

Page 52: Two-Step Problem Solving (continued)

4. a) The coupons total how much money?
   $.35 + $.15 + $.25 + $.65 + $.09 = $1.49

   b) How much will Heidi pay for the groceries with the coupons?
   $45.25 − $1.49 = $43.76

Page 53: Multi-Step Word Problems

Questions will vary. Final answers are listed below.

1. Muriel received $1.89 in change.
2. The family paid $59.51.
3. Chris has 5.6 miles left to go.
4. Millie has $136.91 left.

Page 54: Picture Problems

1. $5.25
2. $18.51
3. $16.50
4. $1.63 + $4.33 + $2.92 = $8.88 total change
5. $17.60
6. $96.75

CHAPTER 6: LIFE-SKILLS MATH

Page 55: Reading Temperatures

a. 100.5°
b. 100.1°
c. 100.9°
d. 100.7°

1. 98.6°
2. 100.1°
3. 2.2°
4. 3.7°

Page 56: Figuring Change

A. $20.00 − $15.50 = $4.50

1. $5.00 − $3.91 = $1.09
2. $25.00 − $23.49 = $1.51
3. $50.00 − $42.00 = $8.00
4. $5.00 − $1.50 = $3.50

Page 57: Making Change

	Work	$5	$1	25¢	10¢	5¢	1¢
$4.05			4			1	
1. $7.08	$20.00 − $12.92 = $7.08	1	2			1	3
2. $1.52	$5.00 − $3.48 = $1.52		1	2			2
3. $.17					1	1	2
4. $4.16			4		1	1	1
5. $7.59		1	2	2		1	4

Page 58: Money Problems

1. $3.12
2. $251.41
3. $16.35
4. a) jacket and clock
   $.91 less
5. a) Sandwich
   b) Cola
   c) French Fries

Page 59: Keeping Records

1. 15.13 pounds
2. 6.01, 3.50, 3.23, 2.39
3. 3.62 pounds
4. 3.89 pounds
5. $42.88
6. a) Meals; b) Bait; c) Lodging

Page 60: Breaking the Record

1. 52.04
2. .44
3. a) Nancy 12.4
   b) Maria 12.58
   c) Shara 13.2
   d) Carrie 13.86
4. 1.46
5. Yes
   .12 second

CHAPTER 1: MULTIPLICATION

Page 1: Counting Decimal Places

1. 2
2. 1
3. 0
4. 2
5. 2
6. 1
7. 3
8. 0
9. 3
10. 1
11. 3
12. 1
13. 2
14. 2
15. 3
16. 0

Page 2: Decimal Places in Answers

	Top	Bottom	Total
1.	2	+ 2	4
2.	2	+ 1	3
3.	3	+ 1	4
4.	0	+ 2	2
5.	3	+ 2	5
6.	2	+ 0	2
7.	2	+ 1	3
8.	1	+ 2	3

Page 3: Place the Decimal Point

1. 83.4
2. 4.027
3. 1.6425
4. .97
5. 683
6. 9.06
7. .7513
8. 17
9. 2.374
10. 7.4

	Factor	Places	Factor	Places	Product	Places
11.	7.5	1	× .3	1	2.25	2
12.	2.47	2	× .4	1	.988	3
13.	.906	3	× .7	1	.6342	4

14.

.89	2
× 53	0
267	
445	
47.17	2

Page 4: Tenths Times a Whole Number

A. 0.9

	Factor	Places	Factor	Places	Product	Places
B.	0.3	1	× 3	0	0.9	1

Page 4: Tenths Times a Whole Number (continued)

1. a) 1.2
2. a) 18.4
3. a) 113.8
4. a) 973.5

	Factor	Places	Factor	Places	Product	Places
1. b)	0.4	1	× 3	0	1.2	1
2. b)	4.6	1	× 4	0	18.4	1
3. b)	56.9	1	× 2	0	113.8	1
4. b)	324.5	1	× 3	0	973.5	1

Page 5: Hundredths Times a Whole Number

A. .75

	Factor	Places	Factor	Places	Product	Places
B.	0.25	2	× 3	0	0.75	2

1. a) 1.14

	Factor	Places	Factor	Places	Product	Places
1. b)	0.38	2	× 3	0	1.14	2

2. a) 24.20

	Factor	Factor	Product
2. b)	6.05	× 4	24.20

3. a) 184.82
   b) 184.82
4. a) 1420.86
   b) 1420.86

Page 6: Multiplying Without Regrouping

	Factor	Places	Factor	Places	Product	Places
1.	.9	1	× .7	1	.63	2
2.	2.3	1	× .3	1	.69	2
3.	.05	2	× 9	0	.45	2
4.	6.21	2	× .4	1	2.484	3
5.	.403	3	× .3	1	.1209	4
6.	.8	1	× .4	1	.32	2
7.	.534	3	× .2	1	.1068	4
8.	8.23	2	× .2	1	1.646	3
9.	11.2	1	× .4	1	4.48	2

Page 7: Multiplication with Regrouping

1. .120
2. 2.94
3. 39.9
4. 1.32
5. 4.98
6. 2.454
7. .987
8. .3661
9. .2152
10. 39.70
11. 1.704
12. 294.4

Page 8: Zeros as Placeholders

A. .06 2
× .8 1
.048 3

B. .17 2
× .03 2
.0051 4

1. .007 3
× .5 1
.0035 4

2. .34 2
× .02 2
.0068 4

3. .009 3
× 4 0
.036 3

4. .003 3
× .06 2
.00018 5

5. .135 3
× .7 1
.0945 4

6. .26 2
× .009 3
.00234 5

Page 9: Practice Adding Zeros

1. .08 2
× .4 1
.032 3

2. .006 3
× 7 0
.042 3

3. .29 2
× .06 2
.0174 4

4. .026 3
× .8 1
.0208 4

5. .302 3
× .03 2
.00906 5

6. .54 2
× .17 2
.0918 4

7. .027 3
× .05 2
.00135 5

8. .099 3
× .02 2
.00198 5

9. .642 3
× .018 3
.011556 6

10. 1.14 2
× .03 2
.0342 4

11. 9.89 2
× .002 3
.01978 5

12. .25 2
× .35 2
.0875 4

Page 10: Line Up and Multiply

1. 1.4
× 8
11.2

2. 6.7
× 5
33.5

3. 2.6
× 3
7.8

4. 2.3
× .4
.92

5. 4.7
× 5
23.5

6. 8.9
× 6
5.34

7. .84
× .8
.672

8. .12
× .7
.084

9. .45
× .2
.090

Page 11: Longer Number on Top

A. 1.3
× .7
.91

B. 7.9
× .5
3.95

1. 6.2
× 8
49.6

2. 25
× .7
17.5

3. 4.9
× 3
14.7

4. 16
× .4
6.4

5. 2.7
× .6
1.62

6. 8.6
× .1
.86

7. 1.8
× .9
1.62

8. 17
× .4
6.8

9. 3.3
× .5
1.65

10. 4.8
× 6
28.8

11. 7.9
× .3
2.37

12. 6.9
× .2
1.38

Page 12: Multiplying by Hundredths

A. .56
× .63
168
3360
.3528

B. .45
× .14
180
450
.0630

C. 7.2
× .38
576
2160
2.736

1. .5655
2. 279.3
3. 15.12
4. .0925
5. .817
6. 34.08
7. 9.114
8. 1.608
9. 56.42

Page 13: Practice Helps

1.	5.	9.
4.35	.701	9.45
× .38	× .43	× 5.6
3480	2103	5670
13050	28040	47250
1.6530	.30143	52.920

2. 5.396
3. 18.565
4. .40426
6. .02888
7. 406.00
8. .03151
10. 382.28
11. 1.6544
12. 140.55

Page 14: Put It All Together

1. 2.44
2. 5.7
3. .738
4. 1.050
5. 256.23
6. 1.242
7. 33.75
8. .00584
9. .04515
10. 330.33
11. .054
12. 16.89

Page 15: Multiplying by 10, 100, and 1,000

1. 59.6
2. 6. or 6
3. .4
4. 23.84
5. 8.03
6. 326. or 326
7. 68. or 68
8. 435. or 435
9. 12.3
10. 602.7
11. 42. or 42
12. 6,019. or 6,019
13. 1,093. or 1,093
14. 721. or 721
15. 3,224. or 3,224

Page 16: Practice Your Skills

1. 2,400
2. 10
3. 8,270
4. 914.5
5. 425
6. .7
7. 90
8. 34,720
9. 88,120
10. 702.35
11. 7.07
12. 860
13. 33.0
14. 6,260
15. 101.0

Page 17: Multiplication Review

1. 2

2.		3.	
.06	2	5.43	2
× 8.4	1	× .7	1
	3	3.801	3

Page 17: Multiplication Review (continued)

4. a) 156.3

b)

52.1	1
× 3	0
156.3	1

5. .0912
6. 67.28

7.

.002	3
× .09	2
.00018	5

8.

.16	2
× .04	2
.0064	4

9. 1.4
10. 34.4
11. .432
12. 2.116
13. 19.44
14. 1.6544
15. 864.3
16. 63.4

CHAPTER 2: MULTIPLICATION PROBLEM SOLVING

Page 18: Number Sentences

1. $1.35 × 5 = $6.75; 6.75
2. $3.45 × 7 = $24.15; 24.15
3. 4.8 × $.89 = $4.56; 4.56
4. $8.78 × 4.5 = $39.51; 39.51
5. $95.64 × 3 = $286.92; 286.92
6. $12.75 × 5 = $63.75; 63.75

Page 19: Does the Answer Make Sense?

1. 45.5 × 5.5 = 250.25; 250.25
2. 4 × $.75 = $3.00; 3.00
3. 9.6 × $9.45 = $90.72; 90.72
4. $10.75 × 52 = $559.00; 559.00
5. 3.4 × 6 = 20.4; 20.4
6. 2.5 × 3 = 7.5; 7.5

Page 20: Grocery Shopping

1. a) $1.72
   b) $2.55
   c) $1.47
2. a) $3.64
   b) $2.58
   c) $3.18
3. $15.14

CHAPTER 3: DIVISION

Page 21: Understanding Division Words

A. quotient
B. dividend
C. divisor
1. 2
2. 8
3. 16
4. 24
5. 4
6. 6
7. 36
8. 4
9. 9
10. 8
11. 48
12. 6
13. 63
14. 7
15. 9
16. 35
17. 5
18. 7

Page 22: Dividing a Decimal by a Whole Number

1. $.25

$.25
2)$.50
4
10
10
0

2. $.25

$.25
3)$.75
6
15
15
0

3. $.10

$.10
4)$.40
4
00
0
0

4. $.16

$.16
5)$.80
5
30
30
0

Page 23: Move the Decimal Straight Up

1. .3
2. .3
3. 1.5
4. .308

5.
.46
7)3.22
2 8
42
42
0

6. .5
7. .108
8. .6

9.
8.7
4)34.8
32
2 8
2 8
0

10. .47
11. .67
12. .136

Page 24: Zeros as Placeholders

1. .02
2. .005
3. .06
4. .09
5. .001

6.
.007
8).056
0
56
56
0

7. .07
8. .004
9. .002
10. .07

11.
.03
7).21
0
21
21
0

12. .09
13. .004
14. .03
15. .08

Page 25: Dividing by a Decimal

Dividing by a decimal.

Page 26: Place the Decimals

1.
1 4 . 6
.3)4 . 3 8
×10 ×10

2.
.65
.96).6240
×100 ×100

3.
.32
4.9)1.568
×10 ×10

4.
.6
.175).1050
×1000 ×1000

5.
1.8
.9)1.6 2
×10 ×10

6.
5.7
.58)3.306
×100 ×100

7.
45.6
.04)1.824
×100 ×100

8.
78.9
.007).5523
×1000 ×1000

9.
2.5
1.9)4.75
×10 ×10

Page 27: Dividing by Tenths

1. $3.3\overline{).528}$ = .16
   - 33
   - 198
   - 198
   - 0
2. 6
3. .04
4. 8
5. .92
6. 3.4
7. .82
8. 4.5
9. .27

Page 28: Dividing by Hundredths

1. $.07\overline{)4.41}$ = 63.
   - 4 2
   - 21
   - 21
   - 0
2. 4.1
3. 3
4. 4.1
5. 40.9
6. 14
7. 4.5
8. 36
9. 292.1

Page 29: Dividing by Thousandths

1. 256	4. 1,406	7. 23.1
2. 7.8	5. 139	8. 89
3. 4,701	6. 789	9. 14.7

Page 30: Mixed Practice

1. 7	5. 10.6	9. .13
2. 70.9	6. .08	10. 285.2
3. 32	7. 2.4	11. .47
4. 14.7	8. .8	12. 40

Page 31: Zeros in the Dividend

1. 50	4. 428	7. 85
2. 85	5. 75	8. 20
3. 245	6. 800	9. 2,635

Page 32: Whole Numbers Divided by Decimals

1. 5	5. 325	9. 125
2. 30	6. 225	10. 625
3. 75	7. 700	11. 750
4. 2	8. 1,200	12. 12,000

Page 33: Zeros in the Quotient

A. $.8\overline{).016}$ = .02
   - 16
   - 0

B. $56.5\overline{).5085}$ = .009
   - 5085
   - 0

C. $4.7\overline{).141}$ = .03
   - 141
   - 0

1. .008
2. .09
3. .006
4. .007
5. .07
6. .0004
7. .056
8. .0006
9. .05

Page 34: Work for a Zero Remainder

1. $5\overline{)17.0}$ = 3.4
   - 15
   - 2 0
   - 2 0
   - 0
2. .375
3. $4\overline{)5.00}$ = 1.25
   - 4
   - 1 0
   - 8
   - 20
   - 20
   - 0
4. .875
5. $4\overline{)7.00}$ = 1.75
   - 4
   - 30
   - 28
   - 20
   - 20
   - 0
6. .625
7. .25
8. .75
9. 1.25
10. 2.75
11. 3.25
12. 6.25

Page 35: Dividing by 10, 100, and 1,000

1. 63.53	6. 9.345	11. .1962
2. .51	7. 4.39	12. .841
3. .208	8. 1.674	13. .3054
4. 4.76	9. .41	14. .623
5. 42.5	10. .9281	15. .480

Page 36: Mastering the Skills

1. .04
2. 2.3
3. .2
4. .7
5. .5
6. 1.01
7. 9
8. 80
9. 150
10. .004
11. 60
12. .4

Page 37: Division Review

1. a) 24
   b) 3
   c) 8
2. .07
3. .006
4. 9.2
5. 4
6. 23.9
7. .26
8. 31
9. 56.8
10. 130
11. 12,000
12. .02
13. 1.25
14. 8.36
15. 4.26

Page 38: Use All Operations

1. 9.53
2. 73
3. 1,047
4. 148
5. 368
6. .18
7. 82
8. 64.952
9. 297.18
10. 45.37
11. 798.39
12. 558.44

CHAPTER 4: DIVISION PROBLEM SOLVING

Page 39: Finding Averages

A. 86.6°
1. $1.09
2. 21.63 miles per gallon
3. 85.1°
4. 37.19

Page 40: Does the Answer Make Sense?

1. $14.10 ÷ 6 = $2.35; 2.35
2. $8.75 ÷ 7 = $1.25; 1.25
3. $1.95 ÷ $.39 = 5; 5
4. $8.25 ÷ $.75 = 11; 11
5. $31.50 ÷ 5 = 6.3; 6.3
6. $18.00 ÷ 4.5 = $4.00; 4.00

Page 41: Choose to Multiply or Divide

A. $3.00
B. $.50

Page 41: Choose to Multiply or Divide (continued)

1. divide
   You need to find one part of a total.
2. multiply
   You need to find the total cost of several rolls.
3. divide
   You need to find one part of a total.
4. divide
   You need to find one part of a total.
5. multiply
   You need to find the total cost of several gallons.
6. multiply
   You need to find the total cost of several pounds.

Page 42: Decide to Multiply or Divide

1. 52.3 × 3 = 156.9; 156.9
2. $13.23 ÷ 13.5 = $.98; .98
3. 123.2 ÷ 2.2 = 56; 56
4. $.95 × 4 = $3.80; 3.80
5. $42.75 ÷ $4.75 = 9; 9
6. $12.38 × 3 = $37.14; 37.14
7. $28.16 × 3 = $84.48; 84.48
8. $3.90 ÷ $.78 = 5; 5

Page 43: Think It Through

Answers should be similar to these.

1. a) How much will one belt cost?
   b) How much will 3 belts cost?
2. a) How many sacks of apples?
   b) What will 6 sacks of apples cost?
3. a) What will one gallon of paint cost?
   b) What will 3 gallons of paint cost?
4. a) What will 1.25 pounds of beef cost?
   b) What will 5 pounds of beef cost?

Page 43: Think It Through (continued)

5. a) How many people in each van?
   b) What will 32 people pay altogether?

6. a) How many boys and girls bought tickets?
   b) How much did the 40 boys and girls pay for their tickets?

Page 44: Write a Question

Answers will vary.

Page 45: Using Symbols

1. <
2. >
3. >
4. =
5. >
6. <
7. =
8. <

Page 46: Choose the Operation

1. multiplication
2. division
3. subtraction
4. addition
5. addition
6. multiplication
7. subtraction
8. addition
9. multiplication
10. division

Page 47: Apply Your Skills

1. 9.5 × 5 = 47.5; 47.5
2. $3.68 + $4.65 = $8.33; 8.33
3. $9.54 ÷ 3 = $3.18; 3.18
4. $984.16 – $544.46 = $439.70; 439.70
5. $479.55 ÷ 5 = $95.91; 95.91
6. 21.5 × 7 = 150.5; 150.5

Page 48: Mixed Practice

1. $91.16 + $100.64 = $191.80; 191.80
2. $18.75 ÷ $3.75 = 5; 5
3. $27.18 + $.92 = $28.10; 28.10
4. $3.15 ÷ $.45 = 7; 7
5. $16.75 + $5.75 = $22.50; 22.50
6. 18.75 × 9 = 168.75; 168.75

Page 49: Review the Operations

1. C
2. A
3. B
4. D
5. $43.47
6. $43.47
7. $125.92 > $82.45
8. $82.45 < $125.92
9. $56.10 ÷ 6 = $9.35
10. $.67 × 15 = $10.05

Page 50: Two-Step Story Problems

1. a) $185.64
   b) $30.94
2. a) $350.00
   b) $324.25
3. a) $32.15
   b) $7.85
4. a) $20.55
   b) $9.45
5. a) $2.25
   b) $2.80

Page 51: More Two-Step Problems

1. $150.72
2. $560.60
3. $5.14
4. $23.84
5. 150
6. $12.00
7. a) $.23
   b) $1.61

Page 52: Multi-Step Word Problems

Answers should be similar to these.

1. Question 1: How much will the adult tickets cost? $27.75

   Question 2: How much will the children's tickets cost? $19.80

   Question 3: How much will the tickets cost altogether? $27.75 + $19.80 = $47.55

2. Question 1: How much did the two pairs of slacks cost? $45

Page 52: Multi-Step Word Problems (continued)

Question 2: How much did the two sport coats cost? $191.70

Question 3: How much did he pay altogether?
$45 + $191.70 + $11.83 = $248.53

3. Question 1: How much did she pay for the ham? $6.75

Question 2: What was her total bill? $8.80

Question 3: How much change did she get back from $20?
$20 − $8.80 = $11.20

4. Question 1: How many hours a week does he work? 32

Question 2: How much does he earn in one week? $296

Question 3: How many weeks will it take him to earn $1,480?
$1,480 ÷ $296 = 5

CHAPTER 5: LIFE-SKILLS MATH

Page 53: Weekly Paycheck Stub

1. $378.94
2. Federal $56.70
   FICA $25.39
   State $15.15
   Total $97.24
3. Medical $5.09
   Union Dues $1.25
   Others $0
   Total $6.34
4. $97.24 + $6.34 = $103.58
5. $378.94 − $103.58 = $275.36

Page 54: Gross Pay and Net Pay

1.

NAME	WEEK ENDING		GROSS PAY		NET PAY
Millie Sanchez	10/13/89		$156.00		$117.06
TAX DEDUCTIONS			OPTIONAL DEDUCTIONS		
FEDERAL	FICA	STATE	MEDICAL	UNION DUES	OTHERS
$23.40	$9.30	$6.24			

2. $38.94
3. $117.06
4. a) $305.25
   b) $244.20
5. a) $272.00
   b) $206.75
6. $6.85

Page 55: Unit Pricing

1. $.15 per ounce
2. $.27 per foot
3. $.21 per ounce
4. $.16 per ounce
5. $.87 per pound
6. $.05 per piece

Page 56: Find the Best Buy

1. Jumbo
2. Jumbo Size / Regular Size

$$\begin{array}{r} 0.12 \\ 24\overline{)2.88} \\ \underline{24} \\ 48 \\ \underline{48} \\ 0 \end{array} \qquad \begin{array}{r} 0.14 \\ 18\overline{)2.52} \\ \underline{18} \\ 72 \\ \underline{72} \\ 0 \end{array}$$

3. $.12 jumbo $.14 regular
4. Jumbo
5. Answers will vary.

Page 57: Comparison Shopping

1. 60-Foot Special
2. 6-pack
3. Store A
4. 26 oz. for $3.12

Page 58: Sales Tax

1. $.05
2. $.11
3. $.17
4. $.02
5. $.20
6. $.03
7. $.18
8. $.06
9. $.10
10. $.15
11. a) $2.14
    b) $.13
    c) $2.27
12. a) $3.29
    b) $.20
    c) $3.49
13. a) $2.70
    b) $.16
    c) $2.86

Page 59: Eating Out

1. $10.54
2. $4.74
3. a) $4.50
   b) $3.92
   c) $2.30
   d) $10.72
   e) $11.15
   f) $8.85
4. $2.36
5. milk shake, french fries, pizza
6. a) $15.75
   b) $4.60
   c) $4.90
   d) $4.50
   e) $29.75
   f) $31.23
   g) $8.77

Page 60: Real-Life Problems

1. a) $2.97
   b) $11.88
2. a) $10.20
   b) $11.59
   c) $4.75
   d) $26.54
3. $1.05
4. a) $.31
   b) $.35
   c) $.33
   d) $.34
5. $86.18
6. $4.19

CHAPTER 1: THE MEANING OF FRACTIONS

Page 1: Denominators

1. $\frac{1}{4}$
2. $\frac{1}{2}$
3. $\frac{1}{8}$
4. $\frac{1}{3}$
5. $\frac{1}{6}$
6. $\frac{1}{5}$
7. $\frac{1}{8}$
8. $\frac{1}{2}$
9. $\frac{1}{8}$
10. $\frac{1}{3}$
11. $\frac{1}{4}$
12. $\frac{1}{4}$
13. $\frac{1}{5}$
14. $\frac{1}{2}$
15. $\frac{1}{3}$
16. $\frac{1}{2}$
17. $\frac{1}{4}$
18. $\frac{1}{4}$
19. $\frac{1}{10}$
20. $\frac{1}{6}$
21. $\frac{1}{4}$
22. $\frac{1}{3}$

Page 2: Numerators

1. $\frac{3}{4}$
2. $\frac{1}{2}$
3. $\frac{2}{5}$
4. $\frac{1}{4}$
5. $\frac{2}{4}$ or $\frac{1}{2}$
6. $\frac{2}{3}$
7. $\frac{3}{4}$
8. $\frac{4}{6}$ or $\frac{2}{3}$
9. $\frac{4}{8}$ or $\frac{1}{2}$
10. $\frac{1}{3}$
11. $\frac{3}{5}$
12. $\frac{3}{6}$ or $\frac{1}{2}$
13. $\frac{3}{4}$
14. $\frac{4}{5}$
15. $\frac{3}{4}$

Page 3: Fractions with Numerators of 1

1. $\frac{1}{5}$ = D
2. $\frac{1}{8}$ = G
3. $\frac{1}{9}$ = H
4. $\frac{1}{7}$ = A
5. $\frac{1}{10}$ = I
6. $\frac{1}{3}$ = F

Page 3: Fractions with Numerators of 1 (continued)

7. $\frac{1}{4}$ = B
8. $\frac{1}{6}$ = E
9. $\frac{1}{2}$ = C

Answers should be similar to these.

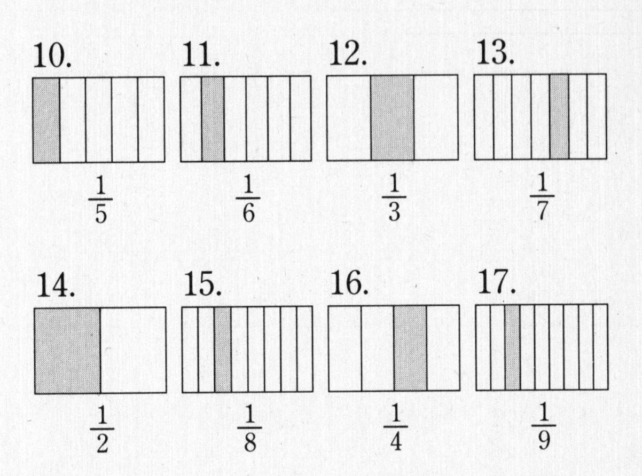

Page 4: Match the Fractions

1. $\frac{2}{3}$ = A
2. $\frac{3}{6}$ = E
3. $\frac{1}{4}$ = F
4. $\frac{5}{6}$ = I
5. $\frac{3}{5}$ = C
6. $\frac{2}{4}$ = B
7. $\frac{5}{8}$ = G
8. $\frac{3}{4}$ = J
9. $\frac{1}{3}$ = D
10. $\frac{1}{2}$ = H

Answers should be similar to these.

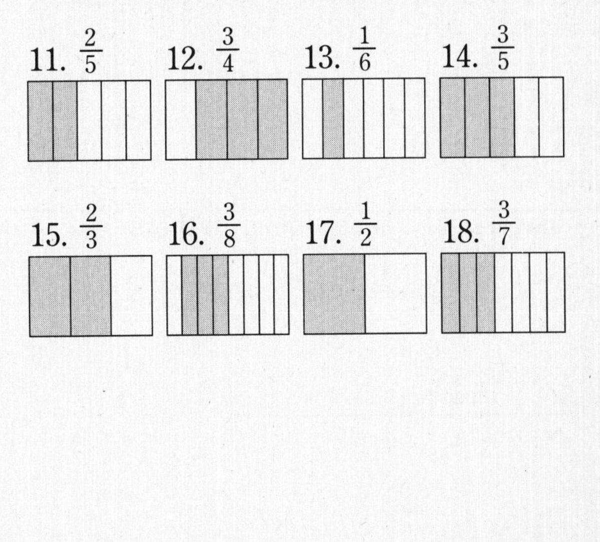

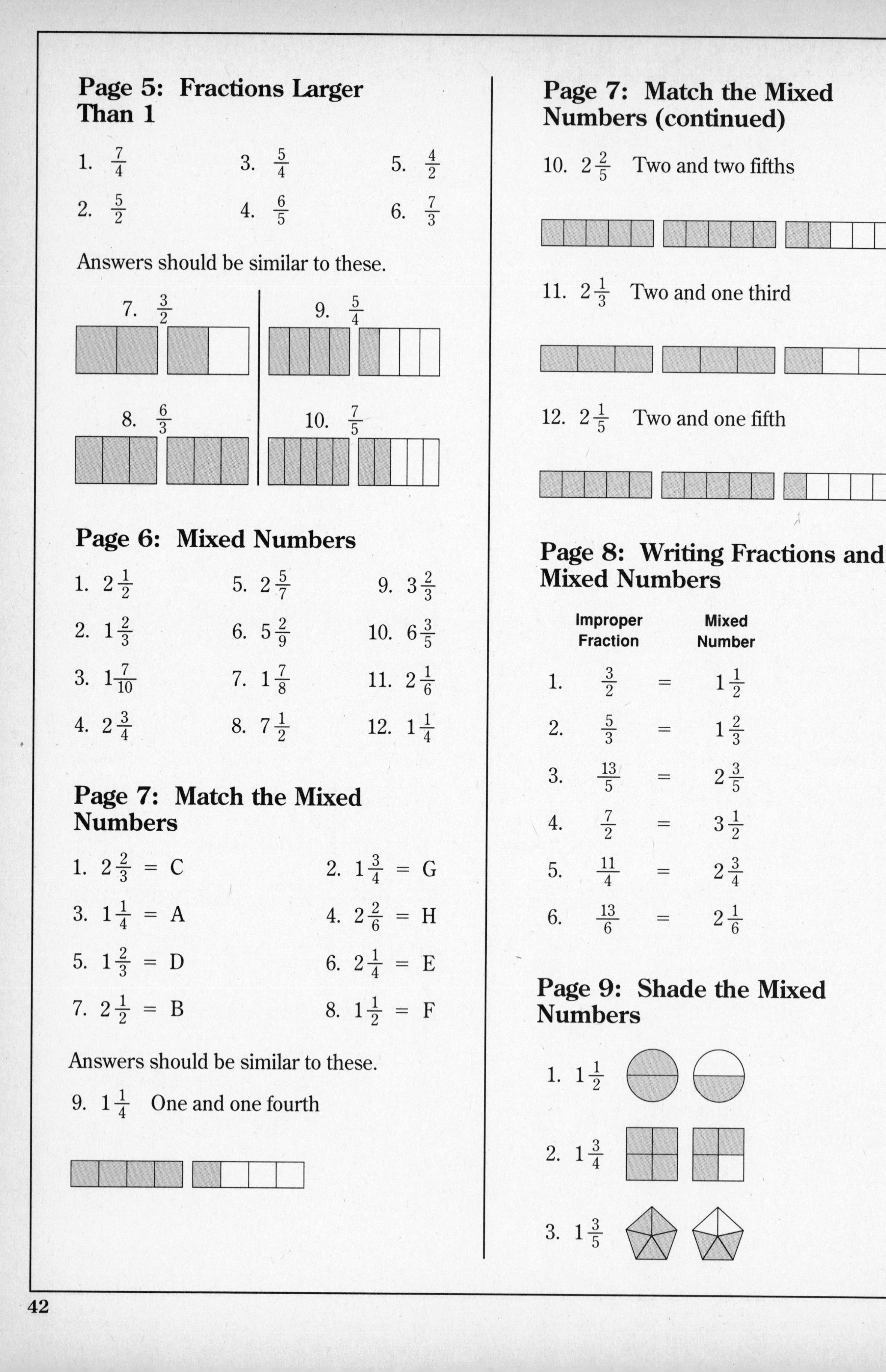

Page 5: Fractions Larger Than 1

1. $\frac{7}{4}$
2. $\frac{5}{2}$
3. $\frac{5}{4}$
4. $\frac{6}{5}$
5. $\frac{4}{2}$
6. $\frac{7}{3}$

Answers should be similar to these.

7. $\frac{3}{2}$
8. $\frac{6}{3}$
9. $\frac{5}{4}$
10. $\frac{7}{5}$

Page 6: Mixed Numbers

1. $2\frac{1}{2}$
2. $1\frac{2}{3}$
3. $1\frac{7}{10}$
4. $2\frac{3}{4}$
5. $2\frac{5}{7}$
6. $5\frac{2}{9}$
7. $1\frac{7}{8}$
8. $7\frac{1}{2}$
9. $3\frac{2}{3}$
10. $6\frac{3}{5}$
11. $2\frac{1}{6}$
12. $1\frac{1}{4}$

Page 7: Match the Mixed Numbers

1. $2\frac{2}{3}$ = C
2. $1\frac{3}{4}$ = G
3. $1\frac{1}{4}$ = A
4. $2\frac{2}{6}$ = H
5. $1\frac{2}{3}$ = D
6. $2\frac{1}{4}$ = E
7. $2\frac{1}{2}$ = B
8. $1\frac{1}{2}$ = F

Answers should be similar to these.

9. $1\frac{1}{4}$ One and one fourth

Page 7: Match the Mixed Numbers (continued)

10. $2\frac{2}{5}$ Two and two fifths

11. $2\frac{1}{3}$ Two and one third

12. $2\frac{1}{5}$ Two and one fifth

Page 8: Writing Fractions and Mixed Numbers

	Improper Fraction		Mixed Number
1.	$\frac{3}{2}$	=	$1\frac{1}{2}$
2.	$\frac{5}{3}$	=	$1\frac{2}{3}$
3.	$\frac{13}{5}$	=	$2\frac{3}{5}$
4.	$\frac{7}{2}$	=	$3\frac{1}{2}$
5.	$\frac{11}{4}$	=	$2\frac{3}{4}$
6.	$\frac{13}{6}$	=	$2\frac{1}{6}$

Page 9: Shade the Mixed Numbers

1. $1\frac{1}{2}$
2. $1\frac{3}{4}$
3. $1\frac{3}{5}$

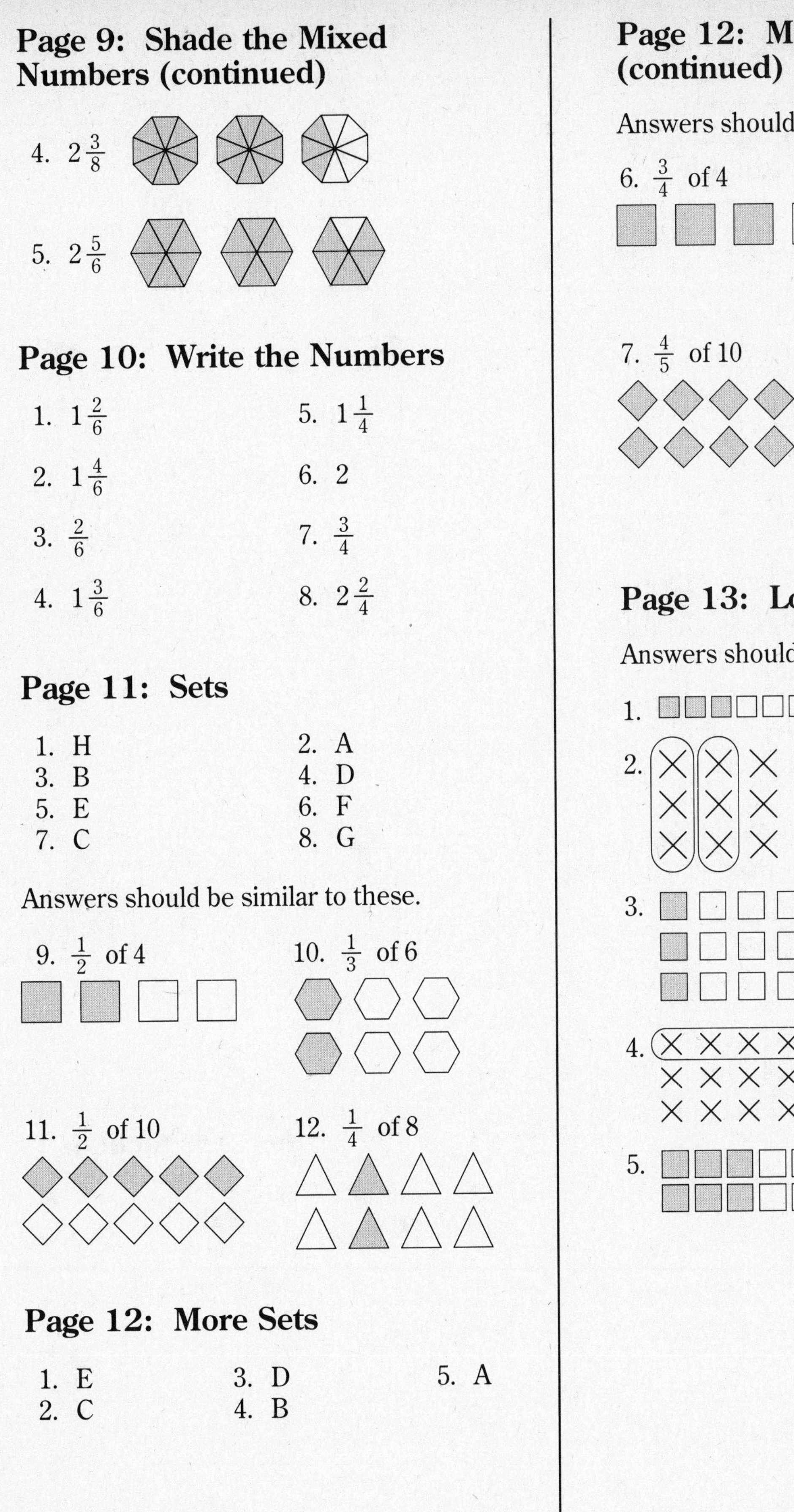

Page 9: Shade the Mixed Numbers (continued)

4. $2\frac{3}{8}$

5. $2\frac{5}{6}$

Page 10: Write the Numbers

1. $1\frac{2}{6}$
2. $1\frac{4}{6}$
3. $\frac{2}{6}$
4. $1\frac{3}{6}$
5. $1\frac{1}{4}$
6. 2
7. $\frac{3}{4}$
8. $2\frac{2}{4}$

Page 11: Sets

1. H
2. A
3. B
4. D
5. E
6. F
7. C
8. G

Answers should be similar to these.

9. $\frac{1}{2}$ of 4

10. $\frac{1}{3}$ of 6

11. $\frac{1}{2}$ of 10

12. $\frac{1}{4}$ of 8

Page 12: More Sets

1. E
2. C
3. D
4. B
5. A

Page 12: More Sets (continued)

Answers should be similar to these.

6. $\frac{3}{4}$ of 4

8. $\frac{2}{3}$ of 6

7. $\frac{4}{5}$ of 10

9. $\frac{3}{4}$ of 12

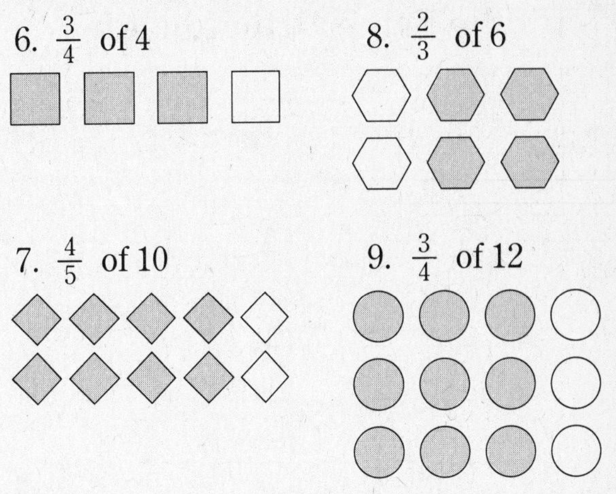

Page 13: Looking at Sets

Answers should be similar to these.

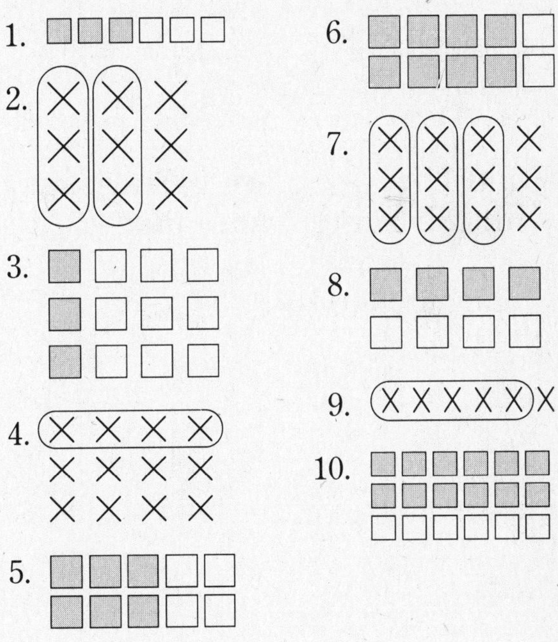

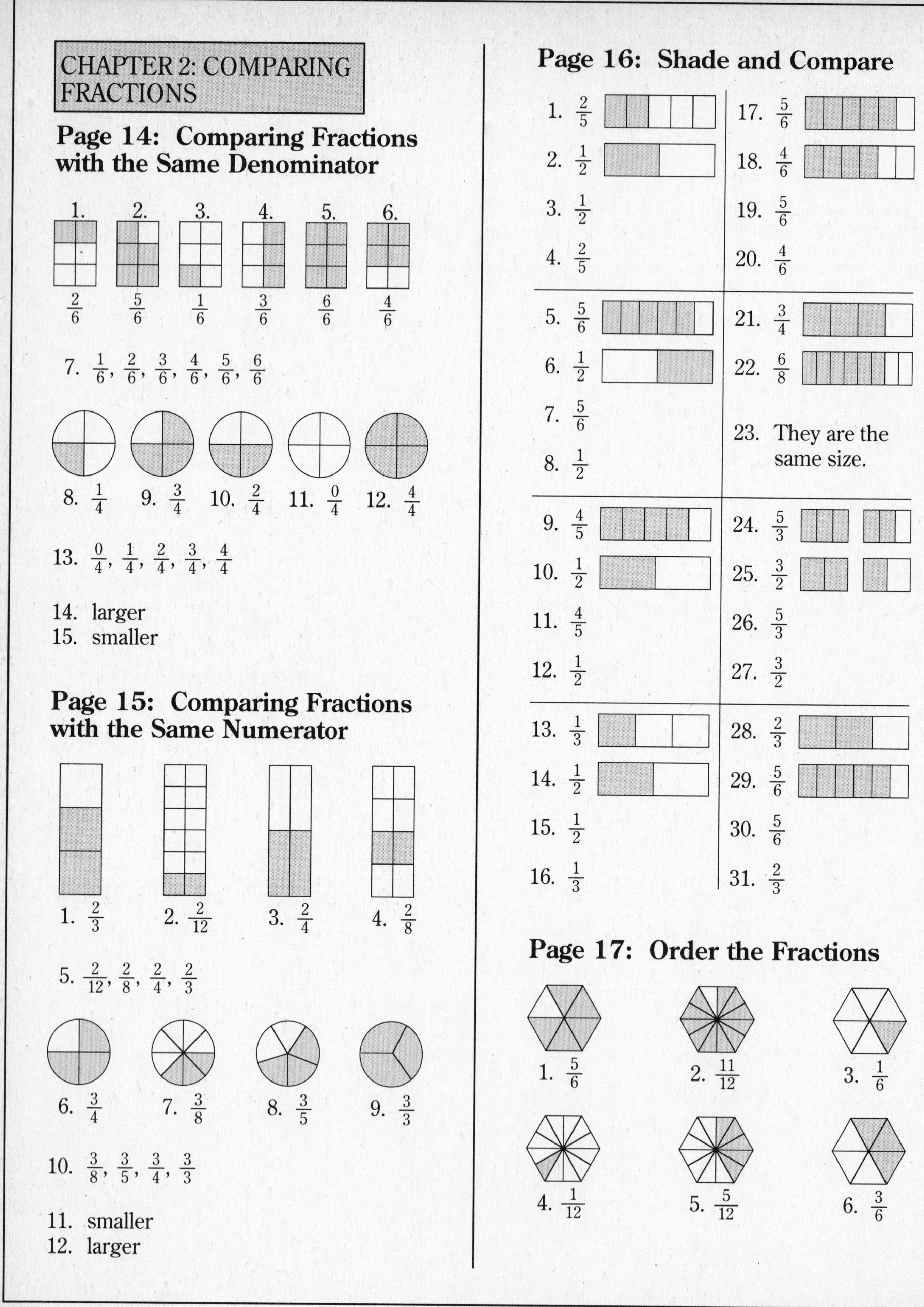

CHAPTER 2: COMPARING FRACTIONS

Page 14: Comparing Fractions with the Same Denominator

1. $\frac{2}{6}$ 2. $\frac{5}{6}$ 3. $\frac{1}{6}$ 4. $\frac{3}{6}$ 5. $\frac{6}{6}$ 6. $\frac{4}{6}$

7. $\frac{1}{6}, \frac{2}{6}, \frac{3}{6}, \frac{4}{6}, \frac{5}{6}, \frac{6}{6}$

8. $\frac{1}{4}$ 9. $\frac{3}{4}$ 10. $\frac{2}{4}$ 11. $\frac{0}{4}$ 12. $\frac{4}{4}$

13. $\frac{0}{4}, \frac{1}{4}, \frac{2}{4}, \frac{3}{4}, \frac{4}{4}$

14. larger
15. smaller

Page 15: Comparing Fractions with the Same Numerator

1. $\frac{2}{3}$ 2. $\frac{2}{12}$ 3. $\frac{2}{4}$ 4. $\frac{2}{8}$

5. $\frac{2}{12}, \frac{2}{8}, \frac{2}{4}, \frac{2}{3}$

6. $\frac{3}{4}$ 7. $\frac{3}{8}$ 8. $\frac{3}{5}$ 9. $\frac{3}{3}$

10. $\frac{3}{8}, \frac{3}{5}, \frac{3}{4}, \frac{3}{3}$

11. smaller
12. larger

Page 16: Shade and Compare

1. $\frac{2}{5}$
2. $\frac{1}{2}$
3. $\frac{1}{2}$
4. $\frac{2}{5}$
5. $\frac{5}{6}$
6. $\frac{1}{2}$
7. $\frac{5}{6}$
8. $\frac{1}{2}$
9. $\frac{4}{5}$
10. $\frac{1}{2}$
11. $\frac{4}{5}$
12. $\frac{1}{2}$
13. $\frac{1}{3}$
14. $\frac{1}{2}$
15. $\frac{1}{2}$
16. $\frac{1}{3}$
17. $\frac{5}{6}$
18. $\frac{4}{6}$
19. $\frac{5}{6}$
20. $\frac{4}{6}$
21. $\frac{3}{4}$
22. $\frac{6}{8}$
23. They are the same size.
24. $\frac{5}{3}$
25. $\frac{3}{2}$
26. $\frac{5}{3}$
27. $\frac{3}{2}$
28. $\frac{2}{3}$
29. $\frac{5}{6}$
30. $\frac{5}{6}$
31. $\frac{2}{3}$

Page 17: Order the Fractions

1. $\frac{5}{6}$ 2. $\frac{11}{12}$ 3. $\frac{1}{6}$

4. $\frac{1}{12}$ 5. $\frac{5}{12}$ 6. $\frac{3}{6}$

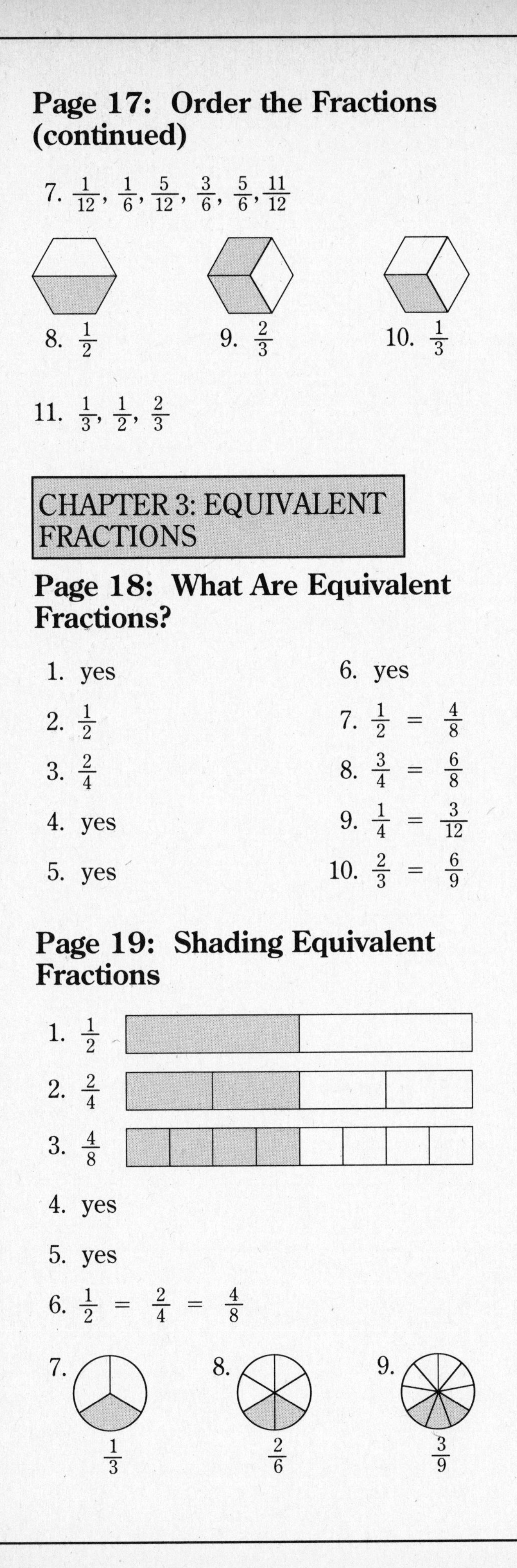

Page 17: Order the Fractions (continued)

7. $\frac{1}{12}, \frac{1}{6}, \frac{5}{12}, \frac{3}{6}, \frac{5}{6}, \frac{11}{12}$

8. $\frac{1}{2}$ 9. $\frac{2}{3}$ 10. $\frac{1}{3}$

11. $\frac{1}{3}, \frac{1}{2}, \frac{2}{3}$

CHAPTER 3: EQUIVALENT FRACTIONS

Page 18: What Are Equivalent Fractions?

1. yes
2. $\frac{1}{2}$
3. $\frac{2}{4}$
4. yes
5. yes
6. yes
7. $\frac{1}{2} = \frac{4}{8}$
8. $\frac{3}{4} = \frac{6}{8}$
9. $\frac{1}{4} = \frac{3}{12}$
10. $\frac{2}{3} = \frac{6}{9}$

Page 19: Shading Equivalent Fractions

1. $\frac{1}{2}$
2. $\frac{2}{4}$
3. $\frac{4}{8}$
4. yes
5. yes
6. $\frac{1}{2} = \frac{2}{4} = \frac{4}{8}$

7. $\frac{1}{3}$ 8. $\frac{2}{6}$ 9. $\frac{3}{9}$

Page 19: Shading Equivalent Fractions (continued)

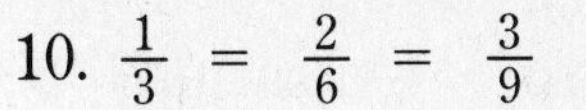

10. $\frac{1}{3} = \frac{2}{6} = \frac{3}{9}$

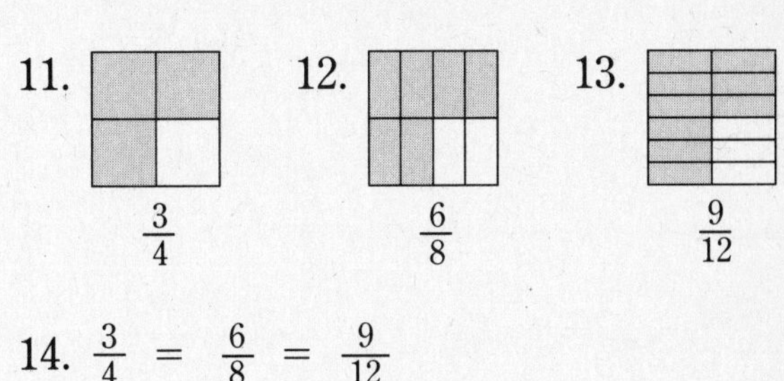

11. $\frac{3}{4}$ 12. $\frac{6}{8}$ 13. $\frac{9}{12}$

14. $\frac{3}{4} = \frac{6}{8} = \frac{9}{12}$

Page 20: Writing Equivalent Fractions

1. $\frac{2}{4}$
2. $\frac{2}{8}$
3. $\frac{2}{16}$
4. $\frac{4}{8}$
5. $\frac{6}{8}$
6. $\frac{2}{4}$
7. $\frac{1}{8}$
8. $\frac{6}{16}$
9. $\frac{8}{8}$
10. $\frac{3}{8}$
11. $\frac{10}{16}$
12. $\frac{1}{4}$
13. $\frac{5}{8}$
14. $\frac{2}{2}$
15. $\frac{12}{16}$
16. $\frac{1}{2}$
17. $\frac{6}{8} = \frac{12}{16}$
18. $\frac{2}{4} = \frac{8}{16}$
19. $\frac{2}{2} = \frac{8}{8}$
20. $\frac{2}{8} = \frac{1}{4}$
21. $\frac{8}{8} = \frac{4}{4}$
22. $\frac{2}{2} = \frac{1}{1}$
23. $\frac{1}{4} = \frac{4}{16}$
24. $\frac{3}{4} = \frac{6}{8}$

Page 21: More Equivalent Fractions

1. $\frac{5}{15}$
2. $\frac{9}{12}$
3. $\frac{20}{32}$
4. $\frac{8}{12}$
5. $\frac{7}{14}$
6. $\frac{12}{20}$
7. $\frac{15}{35}$
8. $\frac{10}{24}$
9. $\frac{15}{18}$
10. $\frac{49}{63}$
11. $\frac{21}{56}$
12. $\frac{28}{48}$

Page 22: Using Fractions Equal to 1

1. $\frac{2}{5} = \frac{2 \times 3}{5 \times 3} = \frac{6}{15}$ so $\frac{2}{5}$ is equal to $\frac{6}{15}$
2. $\frac{3}{4} = \frac{3 \times 4}{4 \times 4} = \frac{12}{16}$ so $\frac{3}{4}$ is equal to $\frac{12}{16}$
3. $\frac{5}{6} = \frac{5 \times 2}{6 \times 2} = \frac{10}{12}$ so $\frac{5}{6}$ is equal to $\frac{10}{12}$
4. $\frac{2}{7} = \frac{2 \times 2}{7 \times 2} = \frac{4}{14}$ so $\frac{2}{7}$ is equal to $\frac{4}{14}$
5. $\frac{1}{8} = \frac{1 \times 3}{8 \times 3} = \frac{3}{24}$ so $\frac{1}{8}$ is equal to $\frac{3}{24}$
6. $\frac{2}{9} = \frac{2 \times 3}{9 \times 3} = \frac{6}{27}$ so $\frac{2}{9}$ is equal to $\frac{6}{27}$

Page 23: Find the Numerators

1. $\frac{1}{4} = \frac{5}{20}$ because $\frac{1 \times 5}{4 \times 5} = \frac{5}{20}$
2. $\frac{3}{5} = \frac{9}{15}$ because $\frac{3 \times 3}{5 \times 3} = \frac{9}{15}$
3. $\frac{4}{7} = \frac{12}{21}$ because $\frac{4 \times 3}{7 \times 3} = \frac{12}{21}$
4. $\frac{1}{3} = \frac{5}{15}$ because $\frac{1 \times 5}{3 \times 5} = \frac{5}{15}$
5. $\frac{5}{6} = \frac{10}{12}$ because $\frac{5 \times 2}{6 \times 2} = \frac{10}{12}$
6. $\frac{2}{9} = \frac{8}{36}$ because $\frac{2 \times 4}{9 \times 4} = \frac{8}{36}$
7. $\frac{2}{3} = \frac{12}{18}$ because $\frac{2 \times 6}{3 \times 6} = \frac{12}{18}$
8. $\frac{3}{4} = \frac{36}{48}$ because $\frac{3 \times 12}{4 \times 12} = \frac{36}{48}$

Page 24: Practice

1. $\frac{1}{3} = \frac{4}{12}$
2. $\frac{3}{4} = \frac{9}{12}$
3. $\frac{1}{2} = \frac{2}{4}$
4. $\frac{2}{3} = \frac{6}{9}$
5. $\frac{1}{4} = \frac{4}{16}$
6. $\frac{1}{5} = \frac{9}{45}$
7. $\frac{1}{3} = \frac{9}{27}$
8. $\frac{1}{7} = \frac{2}{14}$
9. $\frac{2}{5} = \frac{4}{10}$
10. $\frac{1}{6} = \frac{7}{42}$
11. $\frac{7}{10} = \frac{21}{30}$
12. $\frac{5}{7} = \frac{35}{49}$
13. $\frac{7}{12} = \frac{14}{24}$
14. $\frac{11}{50} = \frac{22}{100}$
15. $\frac{9}{20} = \frac{18}{40}$
16. $\frac{7}{15} = \frac{14}{30}$
17. $\frac{11}{14} = \frac{22}{28}$
18. $\frac{7}{8} = \frac{28}{32}$
19. $\frac{5}{9} = \frac{25}{45}$
20. $\frac{1}{4} = \frac{3}{12}$

CHAPTER 4: COMPARISONS

Page 25: Less Than or Greater Than

1. a) $\frac{2}{8}$ b) $\frac{4}{8}$ c) $<$
2. a) $\frac{8}{24}$ b) $\frac{9}{24}$ c) $<$
3. a) $\frac{20}{24}$ b) $\frac{18}{24}$ c) $>$
4. a) $\frac{10}{15}$ b) $\frac{12}{15}$ c) $<$
5. a) $\frac{3}{6}$ b) $\frac{4}{6}$ c) $<$
6. a) $\frac{9}{12}$ b) $\frac{8}{12}$ c) $>$

Page 26: Compare Using the Number Line

1. $<$
2. $=$
3. $>$
4. $>$
5. $<$
6. $=$
7. $>$
8. $<$
9. $=$
10. $>$
11. $>$
12. $<$

Page 27: Connect Tags to Lines

1. $\frac{1}{4}$ C, $\frac{1}{8}$ B, $\frac{5}{8}$ F, $\frac{3}{8}$ D, $\frac{3}{4}$ G $\left(\frac{3}{4} = \frac{6}{8}\right)$, $\frac{7}{8}$ H
2. $\frac{1}{12}$ B, $\frac{1}{4}$ D $\left(\frac{1}{4} = \frac{3}{12}\right)$, $\frac{1}{2}$ G, $\frac{3}{4}$ J, $\frac{2}{3}$ I, $\frac{12}{12}$ M
3. $\frac{1}{10}$ B, $\frac{1}{2}$ F, $\frac{2}{5}$ E, $\frac{3}{5}$ G, $\frac{10}{10}$ K, $\frac{1}{5}$ C

Page 28: Fractions Show Comparisons

1. $\frac{3}{5}$
2. $\frac{3}{4}$
3. $\frac{1}{3}$
4. $\frac{1}{5}$
5. $\frac{1}{2}$
6. $\frac{2}{3}$
7. $\frac{7}{10}$
8. $\frac{1}{8}$

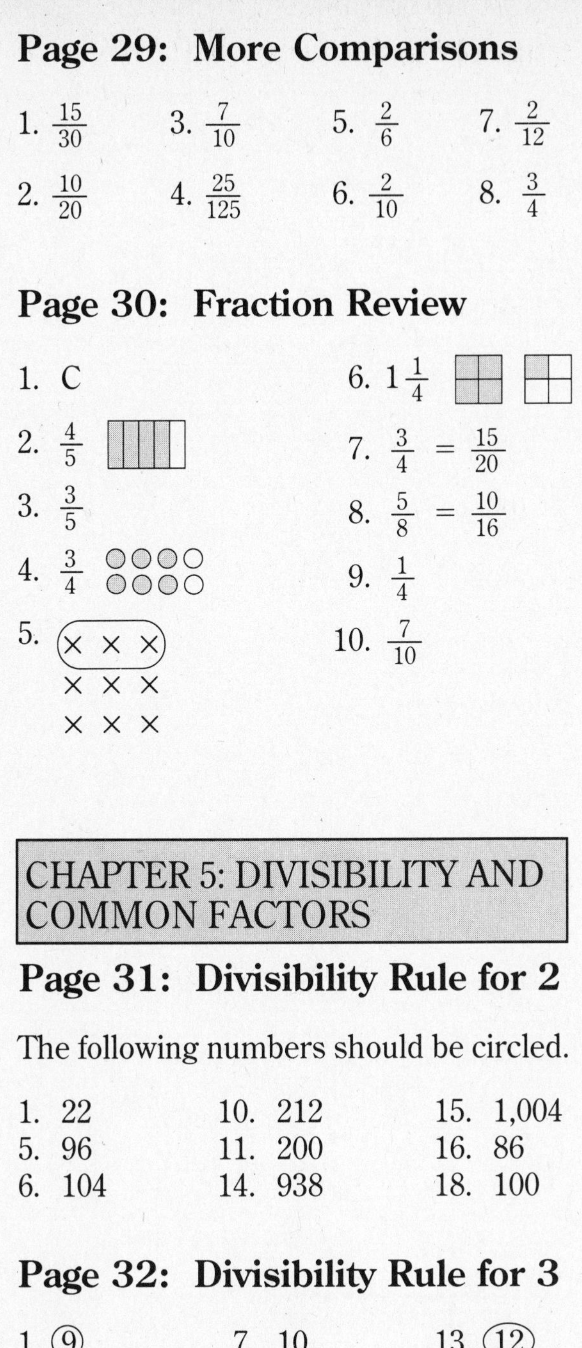

Page 29: More Comparisons

1. $\frac{15}{30}$ 3. $\frac{7}{10}$ 5. $\frac{2}{6}$ 7. $\frac{2}{12}$
2. $\frac{10}{20}$ 4. $\frac{25}{125}$ 6. $\frac{2}{10}$ 8. $\frac{3}{4}$

Page 30: Fraction Review

1. C
2. $\frac{4}{5}$
3. $\frac{3}{5}$
4. $\frac{3}{4}$
5. × × × (circled) × × × × × ×
6. $1\frac{1}{4}$
7. $\frac{3}{4} = \frac{15}{20}$
8. $\frac{5}{8} = \frac{10}{16}$
9. $\frac{1}{4}$
10. $\frac{7}{10}$

CHAPTER 5: DIVISIBILITY AND COMMON FACTORS

Page 31: Divisibility Rule for 2

The following numbers should be circled.

1. 22
5. 96
6. 104
10. 212
11. 200
14. 938
15. 1,004
16. 86
18. 100

Page 32: Divisibility Rule for 3

1. (9)
2. 8
3. (6)
4. 11
5. 11
6. 4
7. 10
8. (12)
9. (3)
10. (6)
11. 10
12. 13
13. (12)
14. (9)
15. (9)
16. 7
17. (15)
18. 8

Page 33: Divisibility Rule for 5

The following numbers should be circled.

2. 55
5. 15
6. 50
8. 35
9. 80
12. 175

Page 33: Divisibility Rule for 5 (continued)

13. 180
14. 105
17. 370
18. 90

Page 34: Divisibility Rule for 10

The following numbers should be circled.

1. 30
8. 70
14. 200
18. 450

Page 35: Divisibility Practice

	Number	Divisible by 2	3	5	10
1.	40	√		√	√
2.	144	√	√		
3.	94	√			
4.	540	√	√	√	√
5.	1,000	√		√	√
6.	29				
7.	45		√	√	
8.	85			√	
9.	342	√	√		
10.	70	√		√	√
11.	100	√		√	√
12.	65			√	
13.	38	√			
14.	153		√		
15.	384	√	√		
16.	89				
17.	420	√	√	√	√
18.	5,546	√			

Page 36: Finding Factors

1. 1, 3, 5, 15
2. 4
3. 1, 2, 4, 5, 10, 20
4. 6

Page 37: Name the Factors

A. 6, 9, 18
B. 6
1. 1, 2, 3, 4, 6, 12
2. 1, 3, 9

Page 37: Name the Factors (continued)

3. 1, 2, 5, 10
4. 1, 2, 4, 7, 14, 28
5. 1, 2, 4, 8
6. 1, 2, 4
7. 1, 7
8. 1, 2, 4, 8, 16
9. 1, 2, 3, 4, 6, 8, 12, 24
10. 1, 2, 3, 5, 6, 10, 15, 30

Page 38: Greatest Common Factor

1. 1, 2, 4
2. 1, 2, 4, 8
3. 1, 2, 4
4. 4
5. 1, 2, 3, 6
6. 1, 3, 9
7. 1, 3
8. 3
9. 1, 2, 4, 8, 16
10. 1, 2, 4, 5, 10, 20
11. 1, 2, 4
12. 4

Page 39: Find the Greatest Common Factor

1. 1, 2, 5, 10
2. 1, 3, 5, 15
3. 1, 5
4. 5
5. 1, 2, 4, 8
6. 1, 2, 3, 4, 6, 12
7. 1, 2, 4
8. 4
9. 1, 2, 7, 14
10. 1, 2, 3, 4, 6, 8, 12, 24
11. 1, 2
12. 2
13. 1, 3, 7, 21
14. 1, 2, 3, 4, 6, 9, 12, 18, 36
15. 1, 3
16. 3

Page 40: Factors with Fractions

A. 1, 2, 3, 6
B. 6
1. 1, 3
2. 3
3. 1, 2, 5, 10
4. 10
5. 1, 2
6. 2
7. 1, 2, 4, 8
8. 8
9. 1, 7
10. 7

Page 41: Apply Your Skills

1. 6
2. 3
3. 2
4. 5
5. 2
6. 7
7. 8
8. 9
9. 2
10. 3
11. 2
12. 6
13. 3
14. 3
15. 3
16. 3
17. 10
18. 5
19. 5
20. 3
21. 9
22. 12
23. 11
24. 7

Page 42: Simplify Fractions

1. GCF = 7; $\frac{2}{7}$
2. GCF = 4; $\frac{1}{3}$
3. GCF = 8; $\frac{3}{4}$
4. GCF = 10; $\frac{2}{3}$
5. GCF = 11; $\frac{1}{3}$
6. GCF = 9; $\frac{2}{5}$
7. GCF = 5; $\frac{5}{6}$
8. GCF = 15; $\frac{2}{3}$
9. GCF = 4; $\frac{8}{9}$
10. GCF = 13; $\frac{2}{3}$

Page 43: Simplify

1. GCF = 6; $\frac{1}{2}$
2. GCF = 3; $\frac{1}{5}$
3. GCF = 2; $\frac{2}{3}$
4. GCF = 5; $\frac{1}{6}$
5. GCF = 2; $\frac{1}{7}$
6. GCF = 7; $\frac{2}{3}$
7. GCF = 8; $\frac{1}{3}$
8. GCF = 9; $\frac{1}{4}$
9. GCF = 2; $\frac{4}{5}$
10. GCF = 3; $\frac{2}{5}$
11. GCF = 2; $\frac{1}{2}$
12. GCF = 6; $\frac{1}{4}$
13. GCF = 3; $\frac{1}{12}$
14. GCF = 3; $\frac{4}{5}$
15. GCF = 3; $\frac{3}{4}$
16. GCF = 3; $\frac{5}{6}$

Page 43: Simplify (continued)

17. GCF = 10; $\frac{2}{3}$
18. GCF = 5; $\frac{5}{8}$
19. GCF = 5; $\frac{1}{10}$
20. GCF = 3; $\frac{3}{5}$
21. GCF = 9; $\frac{1}{2}$
22. GCF = 12; $\frac{1}{3}$
23. GCF = 11; $\frac{1}{2}$
24. GCF = 7; $\frac{2}{5}$

Page 44: Simplest Form

1. simplified
2. $\frac{5}{6}$
3. simplified
4. $\frac{1}{2}$
5. $\frac{2}{3}$
6. simplified
7. simplified
8. $\frac{3}{4}$
9. simplified
10. simplified
11. $\frac{1}{3}$
12. simplified
13. $\frac{4}{5}$
14. simplified
15. simplified
16. simplified
17. $\frac{3}{5}$
18. simplified
19. $\frac{3}{4}$
20. $\frac{1}{1} = 1$
21. $\frac{2}{3}$
22. $\frac{3}{4}$
23. simplified
24. $\frac{5}{6}$

Page 45: Think It Through

1. $\frac{4}{5}$
2. $\frac{1}{3}$
3. $\frac{3}{4}$
4. $\frac{3}{5}$
5. $\frac{1}{3}$
6. $\frac{2}{5}$
7. $\frac{1}{2}$
8. $\frac{1}{2}$
9. $\frac{1}{2}$
10. $\frac{2}{3}$
11. $\frac{2}{3}$
12. $\frac{2}{5}$

Page 46: Lowest Terms

1. $\frac{1}{3}$
2. $\frac{1}{4}$
3. $\frac{1}{3}$
4. $\frac{1}{4}$
5. $\frac{1}{3}$
6. $\frac{2}{3}$

Page 46: Lowest Terms (continued)

7. $\frac{3}{4}$
8. $\frac{1}{2}$
9. $\frac{1}{2}$
10. $\frac{1}{3}$
11. $\frac{1}{3}$
12. $\frac{1}{3}$
13. simplified
14. $\frac{5}{6}$
15. $\frac{1}{2}$
16. $\frac{3}{4}$
17. $\frac{7}{10}$
18. $\frac{1}{2}$
19. $\frac{2}{7}$
20. $\frac{4}{9}$
21. $\frac{5}{7}$
22. simplified
23. $\frac{5}{6}$
24. $\frac{6}{7}$
25. $\frac{5}{8}$
26. $\frac{2}{3}$
27. $\frac{2}{7}$
28. simplified
29. $\frac{1}{3}$
30. $\frac{5}{7}$

CHAPTER 6: FRACTIONS AND DECIMALS

Page 47: Changing Fractions to Decimals

1. $5\overline{)1.0}$ = .2
2. $4\overline{)3.00}$ = .75
3. $8\overline{)3.000}$ = .375
4. $8\overline{)1.000}$ = .125
5. $5\overline{)3.0}$ = .6
6. $8\overline{)7.000}$ = .875

Page 48: More Practice: Fractions to Decimals

1. 1.25
2. 1.2
3. .75
4. .5
5. .625
6. .8
7. 1.375
8. 3.75
9. .25
10. 2.4
11. .875
12. 2.2

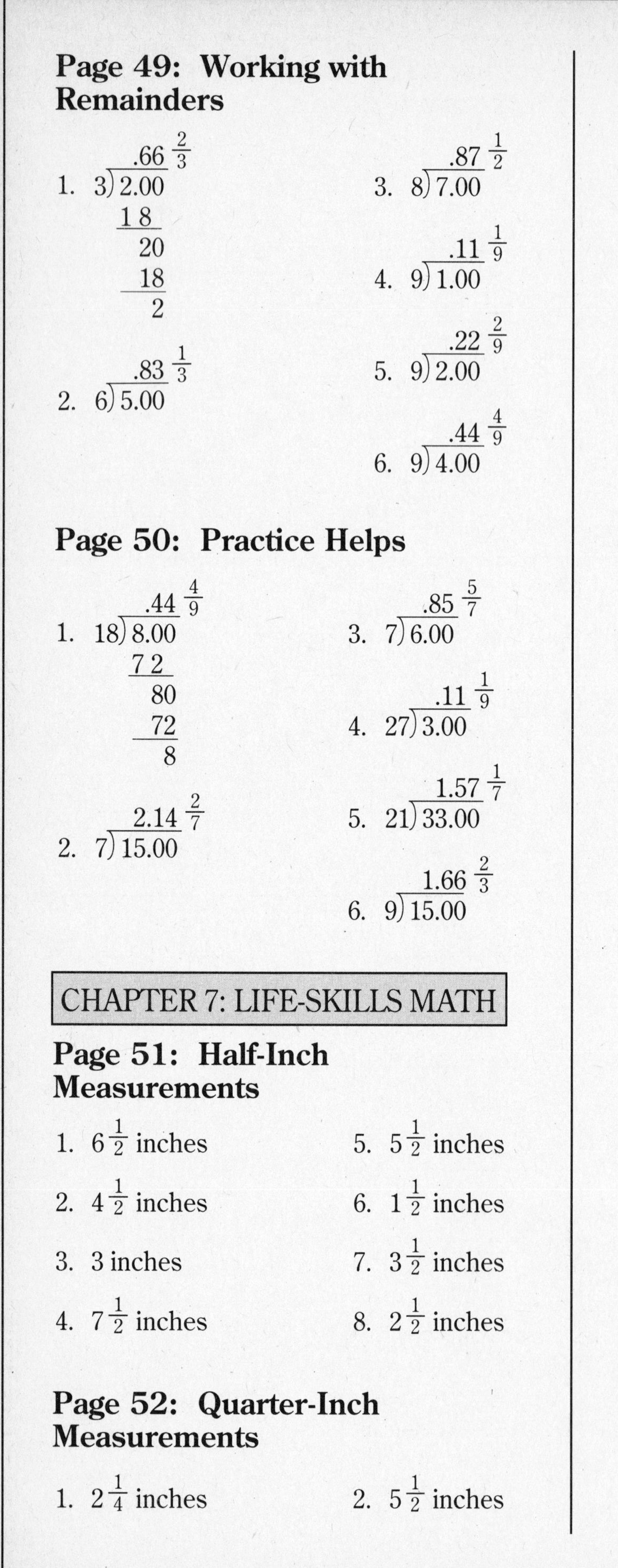

Page 49: Working with Remainders

1. $\begin{array}{r} .66\frac{2}{3} \\ 3\overline{)2.00} \\ \underline{1\,8} \\ 20 \\ \underline{18} \\ 2 \end{array}$

2. $\begin{array}{r} .83\frac{1}{3} \\ 6\overline{)5.00} \end{array}$

3. $\begin{array}{r} .87\frac{1}{2} \\ 8\overline{)7.00} \end{array}$

4. $\begin{array}{r} .11\frac{1}{9} \\ 9\overline{)1.00} \end{array}$

5. $\begin{array}{r} .22\frac{2}{9} \\ 9\overline{)2.00} \end{array}$

6. $\begin{array}{r} .44\frac{4}{9} \\ 9\overline{)4.00} \end{array}$

Page 50: Practice Helps

1. $\begin{array}{r} .44\frac{4}{9} \\ 18\overline{)8.00} \\ \underline{7\,2} \\ 80 \\ \underline{72} \\ 8 \end{array}$

2. $\begin{array}{r} 2.14\frac{2}{7} \\ 7\overline{)15.00} \end{array}$

3. $\begin{array}{r} .85\frac{5}{7} \\ 7\overline{)6.00} \end{array}$

4. $\begin{array}{r} .11\frac{1}{9} \\ 27\overline{)3.00} \end{array}$

5. $\begin{array}{r} 1.57\frac{1}{7} \\ 21\overline{)33.00} \end{array}$

6. $\begin{array}{r} 1.66\frac{2}{3} \\ 9\overline{)15.00} \end{array}$

CHAPTER 7: LIFE-SKILLS MATH

Page 51: Half-Inch Measurements

1. $6\frac{1}{2}$ inches
2. $4\frac{1}{2}$ inches
3. 3 inches
4. $7\frac{1}{2}$ inches
5. $5\frac{1}{2}$ inches
6. $1\frac{1}{2}$ inches
7. $3\frac{1}{2}$ inches
8. $2\frac{1}{2}$ inches

Page 52: Quarter-Inch Measurements

1. $2\frac{1}{4}$ inches
2. $5\frac{1}{2}$ inches

Page 52: Quarter-Inch Measurements (continued)

3. $3\frac{3}{4}$ inches
4. $7\frac{1}{4}$ inches
5. $4\frac{1}{2}$ inches
6. $1\frac{3}{4}$ inches
7. $6\frac{1}{2}$ inches
8. $2\frac{3}{4}$ inches

Page 53: Eighth-Inch Measurements

1. $2\frac{5}{8}$ inches
2. $7\frac{3}{4}$ inches
3. $3\frac{3}{8}$ inches
4. $4\frac{3}{4}$ inches
5. $5\frac{5}{8}$ inches
6. $6\frac{1}{4}$ inches
7. $4\frac{1}{8}$ inches
8. $1\frac{7}{8}$ inches

Page 54: Draw the Measurements

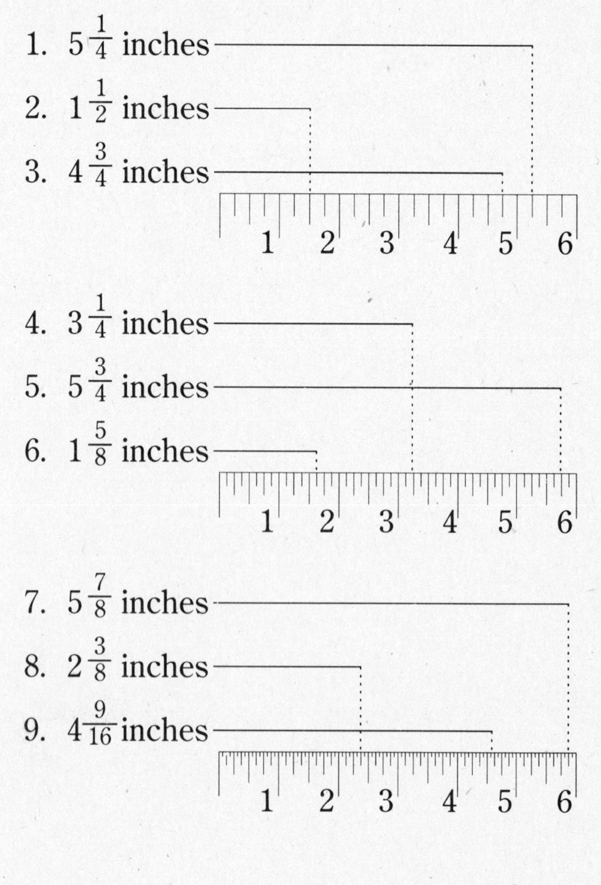

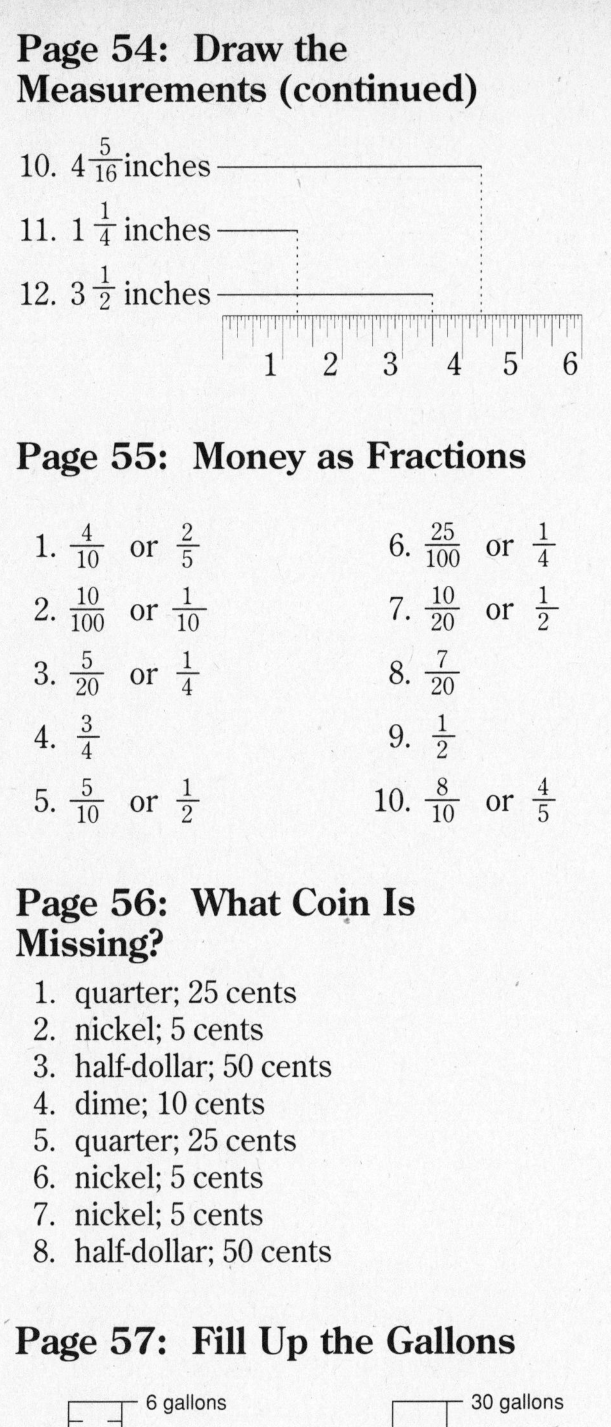

Page 54: Draw the Measurements (continued)

10. $4\frac{5}{16}$ inches
11. $1\frac{1}{4}$ inches
12. $3\frac{1}{2}$ inches

Page 55: Money as Fractions

1. $\frac{4}{10}$ or $\frac{2}{5}$
2. $\frac{10}{100}$ or $\frac{1}{10}$
3. $\frac{5}{20}$ or $\frac{1}{4}$
4. $\frac{3}{4}$
5. $\frac{5}{10}$ or $\frac{1}{2}$
6. $\frac{25}{100}$ or $\frac{1}{4}$
7. $\frac{10}{20}$ or $\frac{1}{2}$
8. $\frac{7}{20}$
9. $\frac{1}{2}$
10. $\frac{8}{10}$ or $\frac{4}{5}$

Page 56: What Coin Is Missing?

1. quarter; 25 cents
2. nickel; 5 cents
3. half-dollar; 50 cents
4. dime; 10 cents
5. quarter; 25 cents
6. nickel; 5 cents
7. nickel; 5 cents
8. half-dollar; 50 cents

Page 57: Fill Up the Gallons

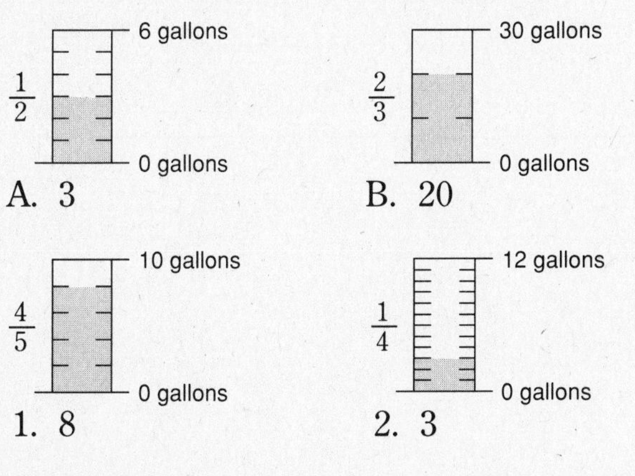

A. 3
B. 20
1. 8
2. 3

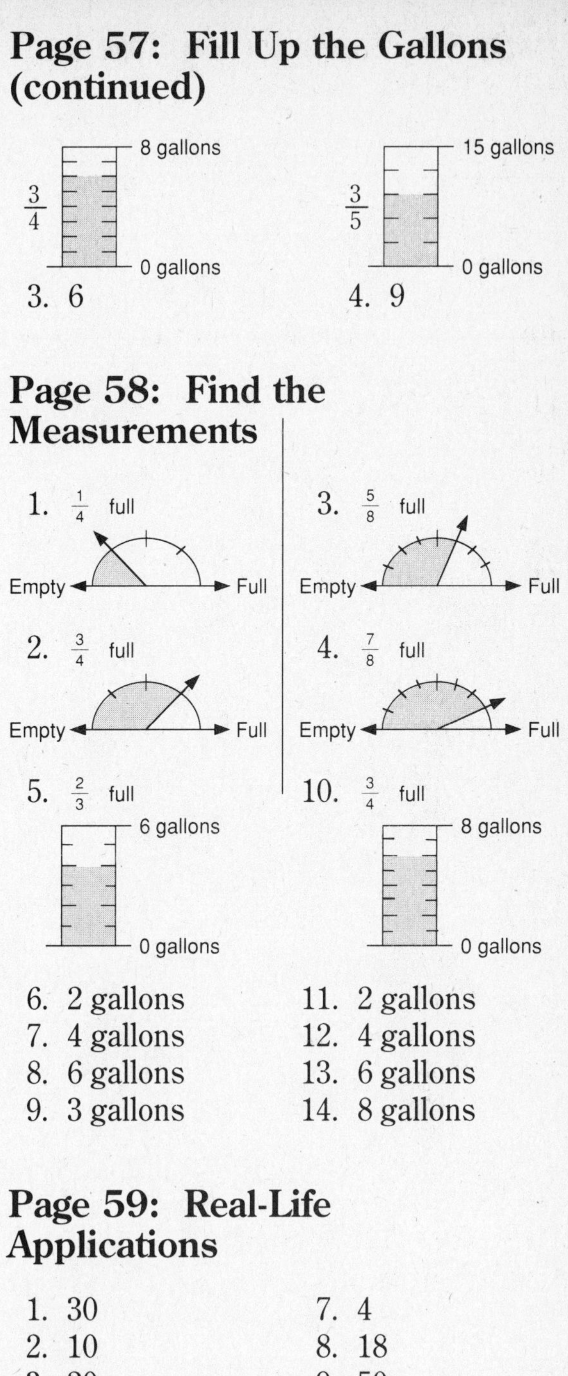

Page 57: Fill Up the Gallons (continued)

3. 6
4. 9

Page 58: Find the Measurements

6. 2 gallons
7. 4 gallons
8. 6 gallons
9. 3 gallons
11. 2 gallons
12. 4 gallons
13. 6 gallons
14. 8 gallons

Page 59: Real-Life Applications

1. 30
2. 10
3. 20
4. 75
5. 6
6. 3
7. 4
8. 18
9. 50
10. 25
11. 75
12. 10

Page 60: Review

1. 10 minutes
2. 16 months
3. a) 3 oranges
   b) 1 gallon
   c) 4 cups
4. $\frac{50}{100}$ or $\frac{1}{2}$
5. $\frac{2}{3}$
6. $\frac{5}{6}$

Page 60: Review (continued)

7. $\frac{1}{7}$
8. $\frac{2}{5}$
9. $\frac{4}{5}$
10. $\frac{3}{5}$
11. $\frac{1}{2}$
12. $\frac{8}{15}$
13. .875
14. .8 or .80
15. 1.25

CHAPTER 1: ADDITION

Page 1: Compare and Add

1. $\frac{3}{9}$ 2. $\frac{4}{9}$ 3. $\frac{7}{9}$

Page 2: Add the Fractions

1. $\frac{1}{4} + \frac{1}{4} = \frac{2}{4}$ or $\frac{1}{2}$
2. $\frac{3}{4} + \frac{1}{4} = \frac{4}{4}$ or 1
3. $\frac{2}{8} + \frac{3}{8} = \frac{5}{8}$
4. $\frac{3}{8} + \frac{4}{8} = \frac{7}{8}$
5. $\frac{6}{16} + \frac{6}{16} = \frac{12}{16}$ or $\frac{3}{4}$
6. $\frac{6}{16} + \frac{5}{16} = \frac{11}{16}$

Page 3: Like Denominators

1. $\frac{6}{7}$
2. $\frac{11}{14}$
3. $\frac{5}{6}$
4. $\frac{9}{11}$
5. $\frac{14}{17}$
6. $\frac{13}{15}$
7. $\frac{11}{15}$
8. $\frac{11}{18}$
9. $\frac{3}{13}$
10. $\frac{9}{11}$

Page 4: Add and Simplify

1. $\frac{2}{4} = \frac{1}{2}$
2. $\frac{5}{10} = \frac{1}{2}$
3. $\frac{9}{12} = \frac{3}{4}$
4. $\frac{10}{15} = \frac{2}{3}$
5. $\frac{15}{18} = \frac{5}{6}$
6. $\frac{15}{21} = \frac{5}{7}$
7. $\frac{5}{15} = \frac{1}{3}$
8. $\frac{6}{9} = \frac{2}{3}$
9. $\frac{15}{20} = \frac{3}{4}$

Page 5: Adding Mixed Numbers

1. $12\frac{3}{9} = 12\frac{1}{3}$
2. $8\frac{6}{8} = 8\frac{3}{4}$

Page 5: Adding Mixed Numbers (continued)

3. $11\frac{5}{8}$
4. $9\frac{6}{12} = 9\frac{1}{2}$
5. $11\frac{9}{10}$
6. $7\frac{2}{4} = 7\frac{1}{2}$
7. $13\frac{7}{9}$
8. $10\frac{3}{6} = 10\frac{1}{2}$
9. $21\frac{4}{10} = 21\frac{2}{5}$
10. $6\frac{5}{7}$
11. $25\frac{12}{16} = 25\frac{3}{4}$
12. $11\frac{8}{15}$

Page 6: Fractions Greater Than 1

1. $\frac{7}{4} = 1\frac{3}{4}$
2. $\frac{11}{5} = 2\frac{1}{5}$
3. $\frac{11}{3} = 3\frac{2}{3}$
4. $\frac{19}{8} = 2\frac{3}{8}$

Page 7: Change to Mixed Numbers

1. $2\frac{2}{4} = 2\frac{1}{2}$
2. $1\frac{1}{6}$
3. $3\frac{4}{6} = 3\frac{2}{3}$
4. $1\frac{4}{5}$
5. $4\frac{1}{5}$
6. $6\frac{2}{6} = 6\frac{1}{3}$
7. $2\frac{1}{3}$
8. $2\frac{5}{10} = 2\frac{1}{2}$
9. $2\frac{1}{5}$
10. $6\frac{3}{9} = 6\frac{1}{3}$
11. 6
12. $7\frac{1}{2}$

Page 8: Practice Simplifying

1. $2\frac{4}{3} = 2 + \frac{4}{3}$
   $= 2 + 1\frac{1}{3}$
   $= 3\frac{1}{3}$

Page 8: Practice Simplifying (continued)

2. $5\frac{6}{5} = 5 + \frac{6}{5}$
   $= 5 + 1\frac{1}{5}$
   $= 6\frac{1}{5}$

3. $6\frac{13}{8} = 6 + \frac{13}{8}$
   $= 6 + 1\frac{5}{8}$
   $= 7\frac{5}{8}$

4. $11\frac{13}{6} = 11 + \frac{13}{6}$
   $= 11 + 2\frac{1}{6}$
   $= 13\frac{1}{6}$

5. $8\frac{19}{9} = 8 + \frac{19}{9}$
   $= 8 + 2\frac{1}{9}$
   $= 10\frac{1}{9}$

6. $7\frac{23}{8} = 7 + \frac{23}{8}$
   $= 7 + 2\frac{7}{8}$
   $= 9\frac{7}{8}$

Page 9: Simplify

1. $3\frac{5}{6}$	7. $11\frac{1}{4}$	13. 17
2. $2\frac{5}{12}$	8. $8\frac{5}{7}$	14. $22\frac{2}{3}$
3. $10\frac{3}{4}$	9. $8\frac{1}{3}$	15. $2\frac{4}{5}$
4. 5	10. $9\frac{1}{3}$	16. $21\frac{1}{2}$
5. $4\frac{1}{4}$	11. $7\frac{2}{3}$	17. $24\frac{2}{3}$
6. 8	12. $14\frac{1}{5}$	18. $6\frac{1}{2}$

Page 10: Simplify Your Answers

1. $\frac{7}{5} = 1\frac{2}{5}$	5. $\frac{12}{8} = 1\frac{4}{8} = 1\frac{1}{2}$
2. $\frac{8}{7} = 1\frac{1}{7}$	6. $\frac{16}{12} = 1\frac{4}{12} = 1\frac{1}{3}$
3. $\frac{5}{4} = 1\frac{1}{4}$	7. $\frac{12}{9} = 1\frac{3}{9} = 1\frac{1}{3}$
4. $\frac{7}{6} = 1\frac{1}{6}$	8. $\frac{12}{10} = 1\frac{2}{10} = 1\frac{1}{5}$

Page 11: Simplify When Necessary

1. $\frac{10}{8} = 1\frac{2}{8} = 1\frac{1}{4}$
2. $6\frac{8}{9}$
3. $\frac{9}{6} = 1\frac{3}{6} = 1\frac{1}{2}$
4. $6\frac{4}{16} = 6\frac{1}{4}$
5. $8\frac{10}{14} = 8\frac{5}{7}$
6. $15\frac{1}{4}$
7. $13\frac{7}{11}$
8. $4\frac{10}{15} = 4\frac{2}{3}$
9. $\frac{15}{13} = 1\frac{2}{13}$
10. $15\frac{4}{4} = 16$
11. $\frac{11}{7} = 1\frac{4}{7}$
12. $8\frac{16}{10} = 9\frac{6}{10} = 9\frac{3}{5}$

CHAPTER 2: LOWEST COMMON DENOMINATOR

Page 12: Multiples

1. 8, 16, 24, 32, 40, 48, 56, 64, 72, 80
2. 8, 16, 24, 32, 40, 48, 56, 64, 72, 80
3. 2, 4, 6, 8, 10, 12, 14, 16, 18, 20
4. 3, 6, 9, 12, 15, 18, 21, 24, 27, 30
5. 4, 8, 12, 16, 20, 24, 28, 32, 36, 40
6. 6, 12, 18, 24, 30, 36, 42, 48, 54, 60
7. 9, 18, 27, 36, 45, 54, 63, 72, 81, 90
8. 12, 24, 36, 48, 60, 72, 84, 96, 108, 120

Page 13: Least Common Multiple

1. 4, 8, 12, 16, 20, 24, 28, 32, 36
2. 6, 12, 18, 24, 30, 36, 42, 48, 54
3. 12, 24, 36
4. 12
5. 3, 6, 9, 12, 15, 18, 21, 24, 27, 30
6. 5, 10, 15, 20, 25, 30, 35, 40, 45, 50
7. 15, 30
8. 15
9. 8, 16, 24, 32, 40, 48, 56, 64, 72, 80
10. 10, 20, 30, 40, 50, 60, 70, 80, 90, 100
11. 40, 80
12. 40

Page 14: Lowest Common Denominator

1. 4, 8, 12, 16, 20, 24, 28, 32,
2. 5, 10, 15, 20, 25, 30, 35, 40,
3. 20
4. 6, 12, 18, 24, 30, 36, 42, 48
5. 8, 16, 24, 32, 40, 48, 56, 64,
6. 24
7. 7, 14, 21, 28, 35, 42, 49, 56
8. 4, 8, 12, 16, 20, 24, 28, 32,
9. 28

Page 15: Find the Lowest Common Denominator (LCD)

1. 4, 8, 12
2. 12, 24, 36
3. 12
4. 6, 12, 18
5. 3, 6, 9
6. 6
7. 5, 10, 15
8. 10, 20, 30
9. 10
10. 8, 16, 24, 32
11. 2, 4, 6, 8
12. 8
13. 6, 12, 18
14. 4, 8, 12
15. 12
16. 9, 18, 27
17. 3, 6, 9
18. 9

Page 16: Choose the Best Method

1. method 2
   LCD is 24
2. method 2
   LCD is 15

Page 16: Choose the Best Method (continued)

3. method 1
   LCD is 12
4. method 2
   LCD is 18
5. method 1
   LCD is 9
6. method 1
   LCD is 22

Page 17: Fraction Readiness

1. LCD = 6
   $\frac{2}{3} = \frac{4}{6}$
   $\frac{1}{2} = \frac{3}{6}$
2. LCD = 12
   $\frac{2}{3} = \frac{8}{12}$
   $\frac{3}{4} = \frac{9}{12}$
3. LCD = 10
   $\frac{3}{5} = \frac{6}{10}$
   $\frac{7}{10} = \frac{7}{10}$
4. LCD = 18
   $\frac{8}{9} = \frac{16}{18}$
   $\frac{5}{6} = \frac{15}{18}$
5. LCD = 12
   $\frac{1}{4} = \frac{3}{12}$
   $\frac{5}{6} = \frac{10}{12}$
6. LCD = 24
   $\frac{5}{12} = \frac{10}{24}$
   $\frac{3}{8} = \frac{9}{24}$
7. LCD = 4
   $\frac{1}{2} = \frac{2}{4}$
   $\frac{1}{4} = \frac{1}{4}$
8. LCD = 24
   $\frac{1}{8} = \frac{3}{24}$
   $\frac{5}{6} = \frac{20}{24}$

Page 18: Compare Unlike Fractions

1. $\frac{1}{3} = \frac{4}{12}$
2. $\frac{1}{4} = \frac{3}{12}$
3. $\frac{1}{3} > \frac{1}{4}$
4. $\frac{1}{2} = \frac{3}{6}$
5. $\frac{5}{6} = \frac{5}{6}$
6. $\frac{1}{2} < \frac{5}{6}$
7. $\frac{2}{3} = \frac{10}{15}$
8. $\frac{4}{5} = \frac{12}{15}$
9. $\frac{2}{3} < \frac{4}{5}$
10. $\frac{5}{6} = \frac{5}{6}$
11. $\frac{2}{3} = \frac{4}{6}$
12. $\frac{5}{6} > \frac{2}{3}$
13. $\frac{3}{8} = \frac{3}{8}$
14. $\frac{3}{4} = \frac{6}{8}$

Page 18: Compare Unlike Fractions (continued)

15. $\frac{3}{8} < \frac{3}{4}$

16. $\frac{1}{3} = \frac{8}{24}$

17. $\frac{3}{8} = \frac{9}{24}$

18. $\frac{1}{3} < \frac{3}{8}$

Page 19: Compare Using Symbols

1. <
2. >
3. >
4. <
5. <
6. <
7. >
8. <
9. >
10. >
11. <
12. >

Page 20: Adding with Unlike Denominators

1. $\frac{4}{6} + \frac{1}{6} = \frac{5}{6}$

2. $\frac{6}{10} + \frac{2}{10} = \frac{8}{10} = \frac{4}{5}$

3. $\frac{1}{12} + \frac{9}{12} = \frac{10}{12} = \frac{5}{6}$

4. $1\frac{7}{12}$

5. $\frac{7}{9}$

6. $\frac{4}{5}$

7. $1\frac{5}{21}$

8. $1\frac{1}{4}$

9. $1\frac{3}{20}$

Page 21: Adding Mixed Numbers with Unlike Denominators

1. $7\frac{3}{15} + 3\frac{10}{15} = 10\frac{13}{15}$

2. $2\frac{3}{6} + 5\frac{4}{6} = 7\frac{7}{6} = 8\frac{1}{6}$

Page 21: Adding Mixed Numbers with Unlike Denominators (continued)

3. $3\frac{11}{12} + 4\frac{9}{12} = 7\frac{20}{12} = 8\frac{8}{12} = 8\frac{2}{3}$

4. $6\frac{1}{2}$

5. $16\frac{1}{4}$

6. $9\frac{13}{20}$

7. $4\frac{13}{14}$

8. $20\frac{1}{5}$

9. $28\frac{2}{3}$

Page 22: Mixed Practice

1. $\frac{13}{24}$
2. $9\frac{1}{2}$
3. $12\frac{7}{24}$
4. $17\frac{1}{5}$
5. $28\frac{4}{5}$
6. $33\frac{1}{2}$
7. $1\frac{7}{12}$
8. $37\frac{5}{6}$
9. $26\frac{2}{3}$
10. $30\frac{11}{18}$
11. $8\frac{2}{3}$
12. $42\frac{1}{3}$

Page 23: Adding Three Numbers

1. $3\frac{2}{8} + 1\frac{1}{8} + 5\frac{4}{8} = 9\frac{7}{8}$

2. $1\frac{4}{5}$

3. $13\frac{7}{24}$

4. $1\frac{1}{4}$

5. $18\frac{1}{5}$

6. $1\frac{5}{18}$

7. $19\frac{5}{8}$

8. $14\frac{3}{4}$

9. $10\frac{11}{20}$

Page 24: Addition Review

1. $\frac{11}{13}$
2. $\frac{6}{15} = \frac{2}{5}$
3. $6\frac{2}{3}$
4. $3\frac{1}{4}$
5. $9\frac{1}{6}$
6. $\frac{10}{8} = 1\frac{2}{8} = 1\frac{1}{4}$
7. $10\frac{5}{5} = 11$
8. 4, 8, 12, 16, 20
9. 6, 12, 18, 24, 30
10. 12
11. 4
12. 18
13. $\frac{1}{4} < \frac{2}{5}$
14. $\frac{12}{10} = 1\frac{2}{10} = 1\frac{1}{5}$
15. $16\frac{19}{24}$
16. $\frac{41}{18} = 2\frac{5}{18}$

CHAPTER 3: ADDITION PROBLEM SOLVING

Page 25: Does the Answer Make Sense?

1. $4\frac{5}{8} + 7\frac{1}{3} = 11\frac{23}{24}$; $11\frac{23}{24}$
2. $6\frac{1}{2} + 3\frac{1}{4} = 9\frac{3}{4}$; $9\frac{3}{4}$
3. $1\frac{1}{2} + \frac{3}{4} = 2\frac{1}{4}$; $2\frac{1}{4}$
4. $5\frac{3}{4} + 12\frac{1}{2} = 18\frac{1}{4}$; $18\frac{1}{4}$
5. $1\frac{5}{8} + 2\frac{3}{4} = 3\frac{11}{8} = 4\frac{3}{8}$; $4\frac{3}{8}$
6. $8\frac{1}{2} + 2\frac{1}{2} = 11$; 11

Page 26: Number Sentences

1. $\frac{3}{4} + \frac{1}{2} = 1\frac{1}{4}$; $1\frac{1}{4}$
2. $\frac{1}{3} + \frac{1}{4} = \frac{7}{12}$; $\frac{7}{12}$
3. $1\frac{3}{4} + 2\frac{1}{2} = 4\frac{1}{4}$; $4\frac{1}{4}$
4. $1\frac{1}{2} + \frac{1}{4} = 1\frac{3}{4}$; $1\frac{3}{4}$
5. $\frac{3}{5} + \frac{1}{5} = \frac{4}{5}$; $\frac{4}{5}$
6. $\frac{1}{3} + \frac{1}{5} = \frac{8}{15}$; $\frac{8}{15}$

CHAPTER 4: SUBTRACTION

Page 27: How Much Is Left?

1. $\frac{1}{5}$
2. $\frac{3}{10}$
3. $\frac{1}{6}$
4. subtract

Page 28: Take Away the Fractions

1. $\frac{3}{4} - \frac{2}{4} = \frac{1}{4}$
2. $\frac{7}{8} - \frac{4}{8} = \frac{3}{8}$
3. $\frac{11}{16} - \frac{6}{16} = \frac{5}{16}$

Page 29: Subtracting with Like Denominators

1. $\frac{3}{7}$
2. $\frac{3}{5}$
3. $\frac{1}{4}$
4. $\frac{2}{5}$
5. $\frac{2}{3}$
6. $\frac{3}{7}$
7. $\frac{3}{5}$
8. $\frac{1}{2}$
9. $\frac{1}{6}$
10. $\frac{6}{11}$
11. $\frac{4}{5}$
12. $\frac{5}{7}$

Page 30: Subtracting Mixed Numbers

1. $2\frac{3}{5}$
2. $5\frac{1}{3}$
3. $7\frac{2}{3}$
4. $3\frac{5}{8}$
5. $4\frac{4}{8} = 4\frac{1}{2}$
6. $3\frac{3}{5}$
7. $5\frac{1}{2}$
8. $9\frac{3}{20}$
9. $6\frac{6}{10} = 6\frac{3}{5}$
10. $8\frac{7}{8}$
11. $6\frac{2}{11}$
12. $10\frac{1}{5}$

Page 31: Subtracting Fractions with Unlike Denominators

1. $\frac{1}{15}$
2. $\frac{9}{12} - \frac{8}{12} = \frac{1}{12}$
3. $\frac{5}{6} - \frac{2}{6} = \frac{3}{6} = \frac{1}{2}$
4. $\frac{4}{12} = \frac{1}{3}$
5. $\frac{2}{9}$
6. $\frac{13}{18}$
7. $\frac{1}{8}$
8. $\frac{5}{10} = \frac{1}{2}$
9. $\frac{9}{12} = \frac{3}{4}$

Page 32: Subtracting Mixed Numbers with Unlike Denominators

1. $7\frac{9}{10} - 3\frac{5}{10} = 4\frac{4}{10} = 4\frac{2}{5}$

2. $8\frac{6}{10} - 1\frac{1}{10} = 7\frac{5}{10} = 7\frac{1}{2}$

3. $6\frac{8}{12} - 2\frac{3}{12} = 4\frac{5}{12}$

4. $4\frac{5}{6}$

5. $2\frac{1}{15}$

6. $3\frac{5}{10} = 3\frac{1}{2}$

7. $2\frac{4}{12} = 2\frac{1}{3}$

8. $4\frac{2}{6} = 4\frac{1}{3}$

9. $6\frac{3}{5}$

Page 33: Subtraction Practice

1. $6\frac{5}{6}$
2. $5\frac{4}{6} = 5\frac{2}{3}$
3. $4\frac{3}{24} = 4\frac{1}{8}$
4. $\frac{13}{18}$
5. $\frac{4}{10} = \frac{2}{5}$
6. $9\frac{5}{15} = 9\frac{1}{3}$
7. $\frac{3}{12} = \frac{1}{4}$
8. $3\frac{1}{24}$
9. $3\frac{2}{12} = 3\frac{1}{6}$
10. $\frac{5}{10} = \frac{1}{2}$
11. $2\frac{7}{8}$
12. $5\frac{4}{30} = 5\frac{2}{15}$

Page 34: Renaming Whole Numbers

1. $3\frac{4}{4}$
2. $2\frac{8}{8}$
3. $4\frac{2}{2}$
4. $2\frac{5}{5}$

Page 35: Mastering the Skill

1. $7\frac{3}{3}$
2. $5\frac{8}{8}$
3. $1\frac{9}{9}$
4. $2\frac{6}{6}$

Page 35: Mastering the Skill (continued)

5. $3\frac{7}{7}$
6. $2\frac{4}{4}$
7. $10\frac{5}{5}$
8. $6\frac{9}{9}$
9. $\frac{10}{10}$
10. $5\frac{14}{14}$
11. $12\frac{15}{15}$
12. $1\frac{3}{3}$
13. $4\frac{2}{2}$
14. $16\frac{4}{4}$
15. $3\frac{9}{9}$
16. $8\frac{7}{7}$
17. $7\frac{12}{12}$
18. $5\frac{20}{20}$

Page 36: Rename and Subtract

1. $5\frac{3}{3} - 2\frac{2}{3} = 3\frac{1}{3}$

2. $3\frac{2}{2} - 3\frac{1}{2} = \frac{1}{2}$

3. $1\frac{3}{5}$
4. $1\frac{1}{8}$
5. $2\frac{3}{10}$
6. $7\frac{1}{12}$
7. $6\frac{1}{4}$
8. $5\frac{2}{7}$
9. $3\frac{1}{8}$
10. $3\frac{1}{5}$
11. $\frac{7}{8}$
12. $1\frac{3}{4}$

Page 37: Rename Mixed Numbers

1. $2\frac{5}{4}$
2. $1\frac{4}{3}$
3. $3\frac{3}{2}$
4. $2\frac{11}{6}$

Page 38: Mixed Number Readiness

1. $4\frac{11}{8}$

   $4 + \frac{8}{8} + \frac{3}{8}$

   $4\frac{11}{8}$

2. $6\frac{8}{5}$

   $6 + 1 + \frac{3}{5}$

   $6 + \frac{5}{5} + \frac{3}{5}$

   $6\frac{8}{5}$

Page 38: Mixed Number Readiness (continued)

3. $3\frac{3}{2}$

   $3 + 1 + \frac{1}{2}$

   $3 + \frac{2}{2} + \frac{1}{2}$

   $3\frac{3}{2}$

4. $7\frac{5}{3}$

   $7 + 1 + \frac{2}{3}$

   $7 + \frac{3}{3} + \frac{2}{3}$

   $7\frac{5}{3}$

Page 39: Renaming Shortcut

1. $5\frac{15}{8}$
2. $3\frac{14}{9}$
3. $12\frac{4}{3}$
4. $8\frac{9}{5}$
5. $6\frac{11}{6}$
6. $1\frac{18}{14}$
7. $6\frac{27}{20}$
8. $4\frac{25}{18}$
9. $12\frac{17}{10}$
10. $18\frac{9}{7}$

Page 40: Rename to Subtract

1. $$\begin{array}{r} 6\frac{11}{8} \\ -\ 4\frac{5}{8} \\ \hline 2\frac{6}{8} = 2\frac{3}{4} \end{array}$$

2. $$\begin{array}{r} 5\frac{7}{5} \\ -\ \ \frac{4}{5} \\ \hline 5\frac{3}{5} \end{array}$$

3. $4\frac{2}{6} = 4\frac{1}{3}$
4. $7\frac{5}{6}$
5. $1\frac{4}{7}$
6. $3\frac{7}{10}$
7. $1\frac{2}{4} = 1\frac{1}{2}$
8. $6\frac{4}{5}$
9. $3\frac{6}{9} = 3\frac{2}{3}$

Page 41: Subtract Unlike Mixed Numbers

1. $$\begin{array}{r} 7\frac{1}{4} = 7\frac{2}{8} = 6\frac{10}{8} \\ -\ 5\frac{3}{8} = 5\frac{3}{8} = 5\frac{3}{8} \\ \hline 1\frac{7}{8} \end{array}$$

2. $2\frac{3}{6} = 2\frac{1}{2}$
3. $2\frac{9}{10}$
4. $1\frac{10}{12} = 1\frac{5}{6}$
5. $3\frac{5}{9}$
6. $4\frac{5}{18}$

Page 41: Subtract Unlike Mixed Numbers (continued)

7. $6\frac{8}{10} = 6\frac{4}{5}$
8. $7\frac{5}{6}$

Page 42: Decide to Rename

1. $4\frac{7}{10}$
2. $4\frac{7}{8}$
3. $2\frac{4}{6} = 2\frac{2}{3}$
4. $8\frac{9}{12} = 8\frac{3}{4}$
5. $4\frac{3}{6} = 4\frac{1}{2}$
6. $6\frac{23}{40}$
7. $3\frac{3}{12} = 3\frac{1}{4}$
8. $7\frac{14}{30} = 7\frac{7}{15}$

Page 43: Mixed Practice

1. $\frac{5}{8}$
2. $4\frac{3}{6} = 4\frac{1}{2}$
3. $6\frac{9}{16}$
4. $6\frac{3}{10}$
5. $6\frac{5}{12}$
6. $2\frac{8}{10} = 2\frac{4}{5}$
7. $8\frac{7}{12}$
8. $4\frac{2}{3}$
9. $10\frac{3}{6} = 10\frac{1}{2}$
10. $5\frac{13}{15}$
11. $\frac{5}{12}$
12. $\frac{2}{6} = \frac{1}{3}$

Page 44: Subtraction Review

1. $\frac{2}{13}$
2. $1\frac{1}{11}$
3. $\frac{7}{12}$
4. $2\frac{1}{15}$
5. $6\frac{1}{6}$
6. $6\frac{4}{4}$
7. $1\frac{2}{5}$
8. $1\frac{3}{2}$
9. $1\frac{2}{3}$
10. $4\frac{5}{18}$
11. $1\frac{3}{8}$
12. $2\frac{9}{11}$

Page 45: Mixed Addition and Subtraction

1. $8\frac{19}{14} = 9\frac{5}{14}$
2. $\frac{24}{20} = 1\frac{4}{20} = 1\frac{1}{5}$
3. $8\frac{2}{6} = 8\frac{1}{3}$
4. $12\frac{12}{10} = 13\frac{2}{10} = 13\frac{1}{5}$

Page 45: Mixed Addition and Subtraction (continued)

5. $7\frac{13}{20}$
6. $8\frac{28}{30} = 8\frac{14}{15}$
7. $9\frac{3}{6} = 9\frac{1}{2}$
8. $\frac{7}{12}$
9. $4\frac{3}{6} = 4\frac{1}{2}$
10. $16\frac{20}{12} = 17\frac{8}{12} = 17\frac{2}{3}$
11. $8\frac{5}{15} = 8\frac{1}{3}$
12. $32\frac{29}{18} = 33\frac{11}{18}$

Page 46: Putting It All Together

1. $\frac{3}{8} < \frac{5}{8}$
2. $1\frac{1}{2} = 1\frac{1}{2}$
3. $9\frac{1}{8} > 8\frac{1}{6}$
4. $1\frac{1}{3} > 1\frac{1}{6}$
5. $11\frac{1}{2} > 11\frac{1}{20}$

CHAPTER 5: SUBTRACTION PROBLEM SOLVING

Page 47: Subtraction Word Problems

1. $5\frac{1}{2} - 1\frac{1}{4} = 4\frac{1}{4}; 4\frac{1}{4}$
2. $\frac{3}{4} - \frac{1}{4} = \frac{1}{2}; \frac{1}{2}$
3. $1\frac{1}{4} - \frac{1}{2} = \frac{3}{4}; \frac{3}{4}$
4. $4\frac{1}{3} - 1\frac{1}{2} = 2\frac{5}{6}; 2\frac{5}{6}$
5. $3\frac{3}{4} - 1\frac{1}{2} = 2\frac{1}{4}; 2\frac{1}{4}$
6. $\frac{8}{8} - \frac{5}{8} = \frac{3}{8}; \frac{3}{8}$

Page 48: Writing Number Sentences

1. $15\frac{2}{3} - 4\frac{1}{4} = 11\frac{5}{12}; 11\frac{5}{12}$
2. $7\frac{1}{2} - 4\frac{3}{4} = 2\frac{3}{4}; 2\frac{3}{4}$
3. $3\frac{1}{2} - 1\frac{1}{4} = 2\frac{1}{4}; 2\frac{1}{4}$
4. $20\frac{1}{2} - 3\frac{1}{3} = 17\frac{1}{6}; 17\frac{1}{6}$
5. $1\frac{1}{3} - \frac{3}{4} = \frac{7}{12}; \frac{7}{12}$
6. $8\frac{1}{2} - 6\frac{3}{8} = 2\frac{1}{8}; 2\frac{1}{8}$

Page 49: Use Drawings to Solve Problems

1. $7\frac{5}{6}$ miles
2. $21\frac{3}{4}$ miles

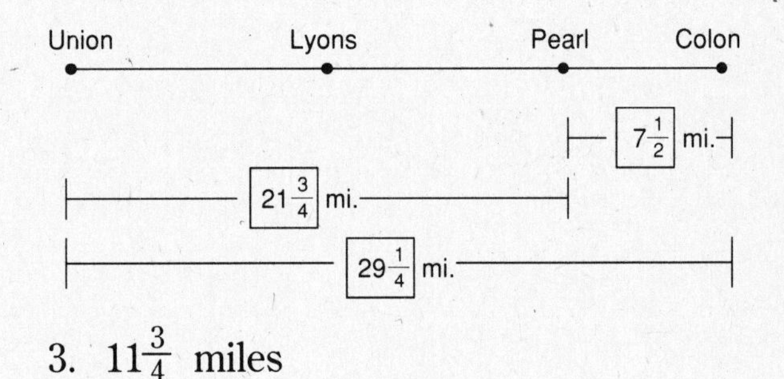

3. $11\frac{3}{4}$ miles

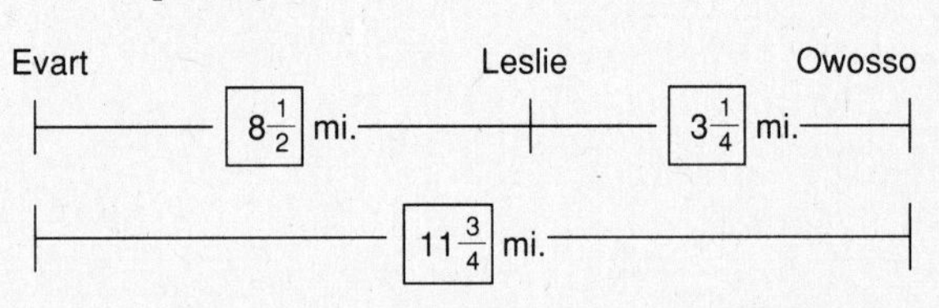

Page 50: Picture Problems

1. $8\frac{3}{4} - 1\frac{1}{2} = 7\frac{1}{4}$
2. $2 - 1\frac{1}{4} = \frac{3}{4}$
3. $\frac{3}{8} + \frac{1}{4} = \frac{5}{8}$
4. $2\frac{1}{2} + \frac{3}{4} = 2\frac{5}{4} = 3\frac{1}{4}$
5. $16\frac{1}{4} - 10\frac{3}{4} = 5\frac{2}{4} = 5\frac{1}{2}$
6. $13\frac{1}{2} + 2\frac{3}{4} = 15\frac{5}{4} = 16\frac{1}{4}$

Page 51: Add or Subtract

1. $1\frac{1}{2} + 2\frac{1}{3} = 3\frac{5}{6}$; $3\frac{5}{6}$
2. $12 - 2\frac{3}{8} = 9\frac{5}{8}$; $9\frac{5}{8}$
3. $12 - 7\frac{3}{4} = 4\frac{1}{4}$; $4\frac{1}{4}$
4. $6\frac{1}{2} + \frac{13}{16} = 6\frac{21}{16}$; $7\frac{5}{16}$

Page 52: Using Symbols

1. $\frac{3}{4} + 1\frac{1}{4} = 2$
2. $2 - 1\frac{1}{4} = \frac{3}{4}$
3. $2 > \frac{3}{4}$
4. $\frac{3}{4} < 2$
5. $\frac{3}{4} \neq 2$
6. $2\frac{1}{2} + 2\frac{3}{4} = 5\frac{1}{4}$
7. $5\frac{1}{4} - 2\frac{3}{4} = 2\frac{1}{2}$
8. $5\frac{1}{4} > 2\frac{1}{2}$
9. $2\frac{1}{2} < 5\frac{1}{4}$
10. $5\frac{1}{4} \neq 2\frac{1}{2}$

Page 53: Write a Question

Questions will vary but should ask for the following items.

1. a) sum of $6\frac{3}{8}$ and $3\frac{1}{4}$

   b) difference of $6\frac{3}{8}$ and $3\frac{1}{4}$

2. a) difference of $3\frac{1}{2}$ and $2\frac{3}{4}$

   b) difference of $3\frac{1}{2}$ and $1\frac{1}{4}$

3. a) difference of $3\frac{1}{2}$ and $1\frac{1}{3}$

   b) sum of $3\frac{1}{2}$ and $1\frac{1}{3}$

4. a) sum of $4\frac{1}{8}$ and $6\frac{1}{2}$

   b) sum of $\frac{7}{8}$ and $\frac{1}{2}$

5. a) difference of $2\frac{1}{4}$ and $1\frac{1}{2}$

   b) sum of $2\frac{1}{4}$ and $1\frac{1}{2}$

6. a) difference of 4 and $1\frac{1}{4}$

   b) difference of 4 and $2\frac{1}{2}$

Page 54: Decide to Add or Subtract

Answers will vary.

Page 55: Mixed Word-Problem Review

1. $\frac{7}{8} - \frac{1}{4} = \frac{5}{8}$; $\frac{5}{8}$
2. $4\frac{5}{8} + 7\frac{9}{16} = 11\frac{19}{16} = 12\frac{3}{16}$; $12\frac{3}{16}$
3. $2\frac{3}{8} - 1\frac{1}{3} = 1\frac{1}{24}$; $1\frac{1}{24}$
4. $4 - 2\frac{1}{8} = 1\frac{7}{8}$; $1\frac{7}{8}$
5. $1\frac{1}{2} + \frac{1}{4} = 1\frac{3}{4}$; $1\frac{3}{4}$
6. $1\frac{1}{3} + \frac{3}{5} = 1\frac{14}{15}$; $1\frac{14}{15}$

CHAPTER 6: LIFE-SKILLS MATH

Page 56: Weighing Food

1. a) 7

   b) Z

   c) $5\frac{1}{2}$ lbs.

   d) $4\frac{3}{4}$ lbs.

2. $3\frac{1}{4} - \frac{3}{8} = 2\frac{7}{8}$
3. $6 - 3\frac{3}{4} = 2\frac{1}{4}$

Page 57: Reading a Measuring Cup

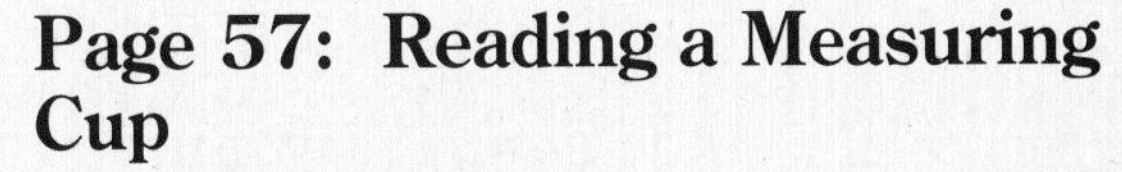

1.

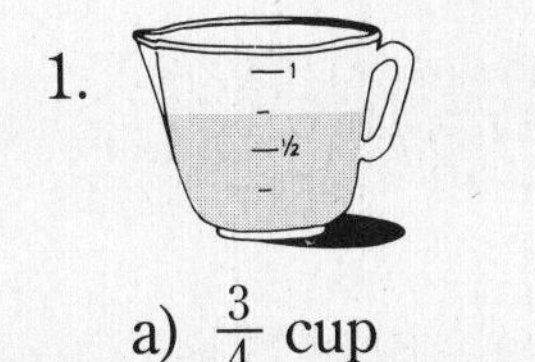

a) $\frac{3}{4}$ cup

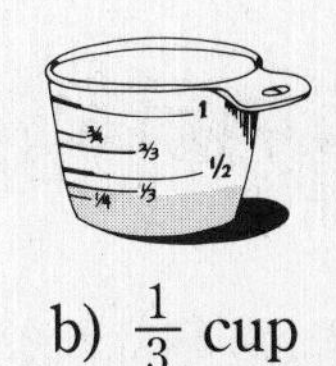

b) $\frac{1}{3}$ cup

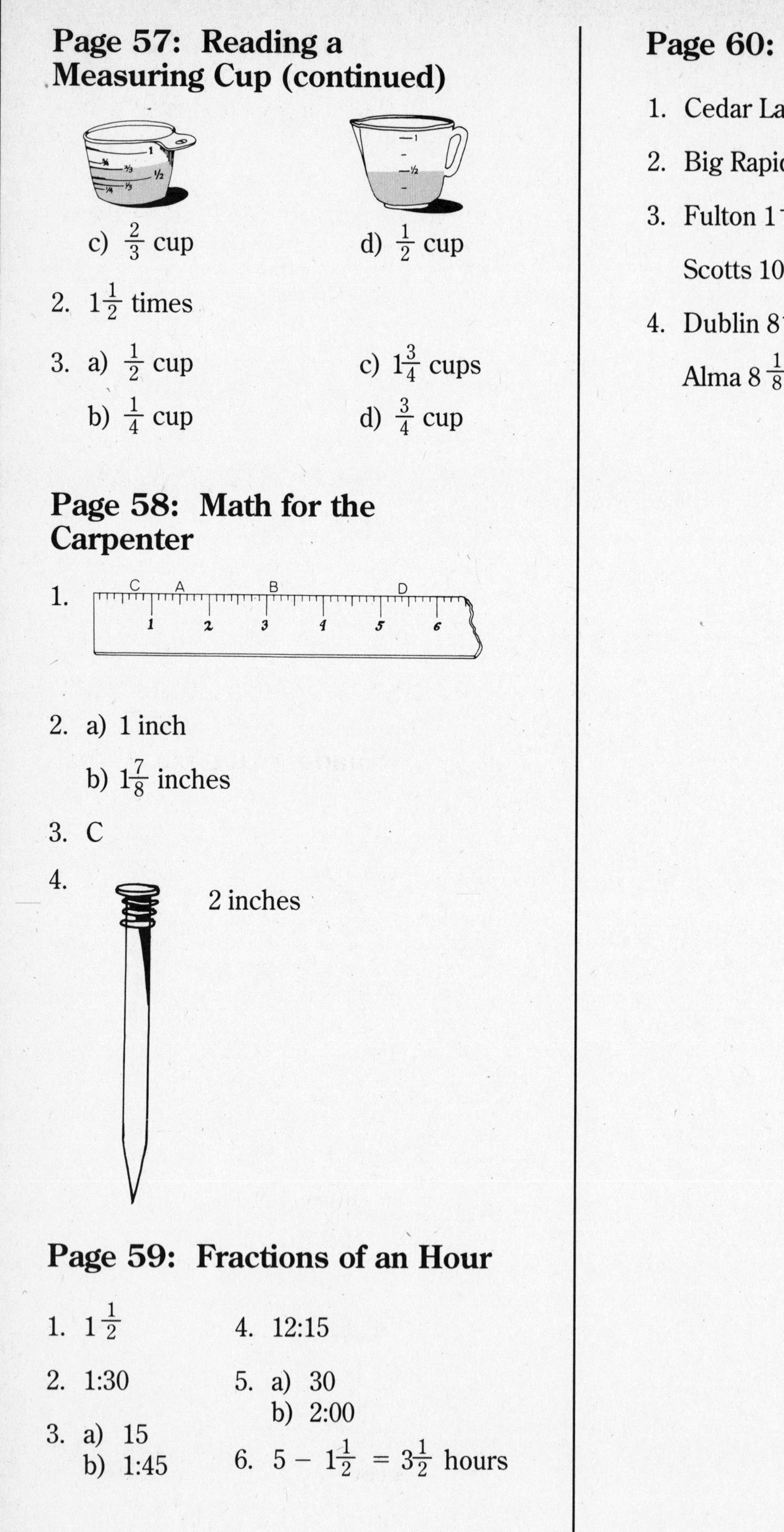

Page 57: Reading a Measuring Cup (continued)

c) $\frac{2}{3}$ cup

d) $\frac{1}{2}$ cup

2. $1\frac{1}{2}$ times

3. a) $\frac{1}{2}$ cup
   b) $\frac{1}{4}$ cup
   c) $1\frac{3}{4}$ cups
   d) $\frac{3}{4}$ cup

Page 58: Math for the Carpenter

1.

2. a) 1 inch
   b) $1\frac{7}{8}$ inches

3. C

4. 2 inches

Page 59: Fractions of an Hour

1. $1\frac{1}{2}$

2. 1:30

3. a) 15
   b) 1:45

4. 12:15

5. a) 30
   b) 2:00

6. $5 - 1\frac{1}{2} = 3\frac{1}{2}$ hours

Page 60: Using Road Signs

1. Cedar Lake $6\frac{3}{4}$ miles

2. Big Rapids $4\frac{1}{4}$ miles

3. Fulton $1\frac{3}{4}$ miles

   Scotts 10 miles

4. Dublin $8\frac{5}{8}$ miles

   Alma $8\frac{1}{8}$ miles

CHAPTER 1: UNDERSTANDING MULTIPLICATION OF FRACTIONS

Page 1: Fractions of a Set

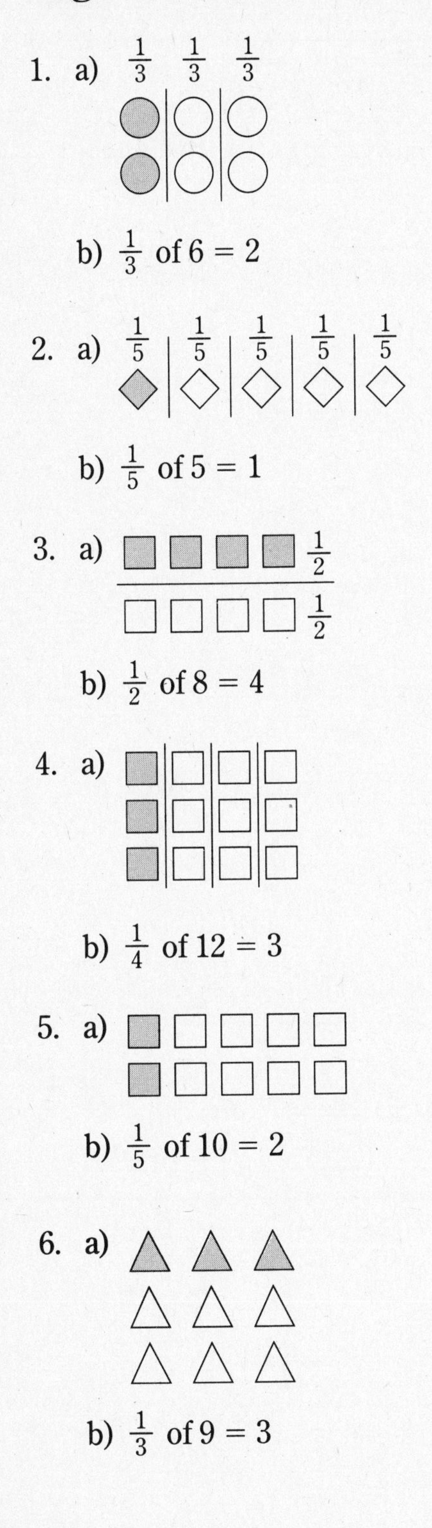

1. a) $\frac{1}{3}$ $\frac{1}{3}$ $\frac{1}{3}$
b) $\frac{1}{3}$ of 6 = 2
2. a) $\frac{1}{5}$ $\frac{1}{5}$ $\frac{1}{5}$ $\frac{1}{5}$ $\frac{1}{5}$
b) $\frac{1}{5}$ of 5 = 1
3. a) $\frac{1}{2}$ $\frac{1}{2}$
b) $\frac{1}{2}$ of 8 = 4
4. a)
b) $\frac{1}{4}$ of 12 = 3
5. a)
b) $\frac{1}{5}$ of 10 = 2
6. a)
b) $\frac{1}{3}$ of 9 = 3

Page 2: Finding One Part of a Set

1. ●●○○ ●●○○
2. yes
3. \$5.00
4. \$10.00
5. \$2.50
6. ■□□
7. yes
8. \$3.00
9. \$10.00
10. \$1.00
11. \$7.00
12. \$2.00

Page 3: Practice Helps

1. 2	6. 8	11. 4	16. 10
2. 4	7. 9	12. 4	17. 6
3. 1	8. 10	13. 4	18. 2
4. 2	9. 8	14. 3	19. 7
5. 1	10. 5	15. 2	20. 1

Page 4: Shade Fractions of the Sets

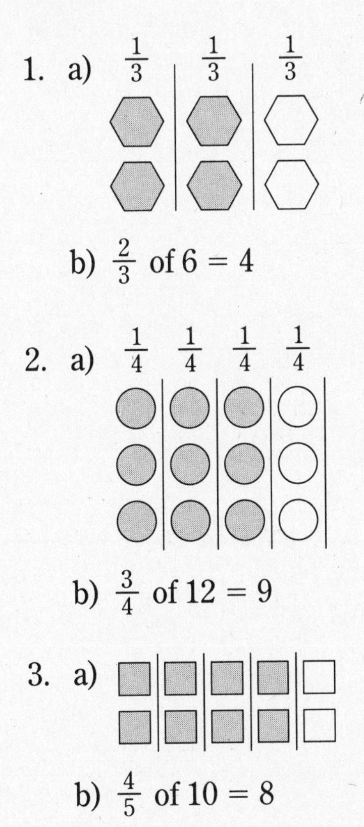

1. a) $\frac{1}{3}$ $\frac{1}{3}$ $\frac{1}{3}$
b) $\frac{2}{3}$ of 6 = 4
2. a) $\frac{1}{4}$ $\frac{1}{4}$ $\frac{1}{4}$ $\frac{1}{4}$
b) $\frac{3}{4}$ of 12 = 9
3. a)
b) $\frac{4}{5}$ of 10 = 8

Page 4: Shade Fractions of the Sets (continued)

4. a) △|△|△|△ (3 of 4 groups shaded)

   b) $\frac{3}{4}$ of 8 = 6

5. a) ◇|◇|◇|◇|◇ (3 of 5 groups shaded)

   b) $\frac{3}{5}$ of 10 = 6

6. a) ⬡|⬡|⬡|⬡ (3 of 4 shaded)

   b) $\frac{3}{4}$ of 4 = 3

Page 5: Finding a Fraction of a Set

1. a) $\frac{1}{4}$ $\frac{1}{4}$ $\frac{1}{4}$ $\frac{1}{4}$

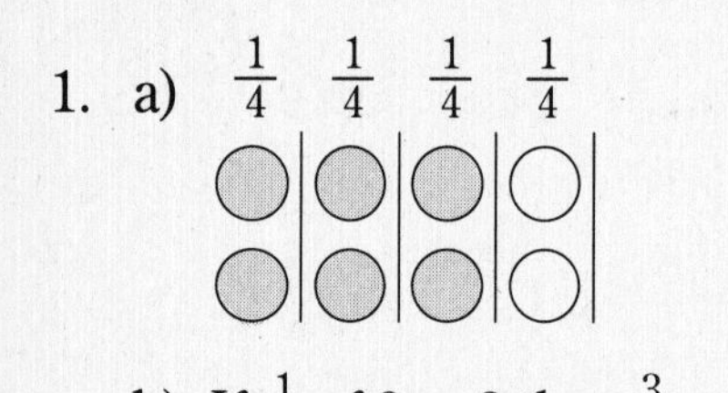

   b) If $\frac{1}{4}$ of 8 = 2 then $\frac{3}{4}$ of 8 = 6

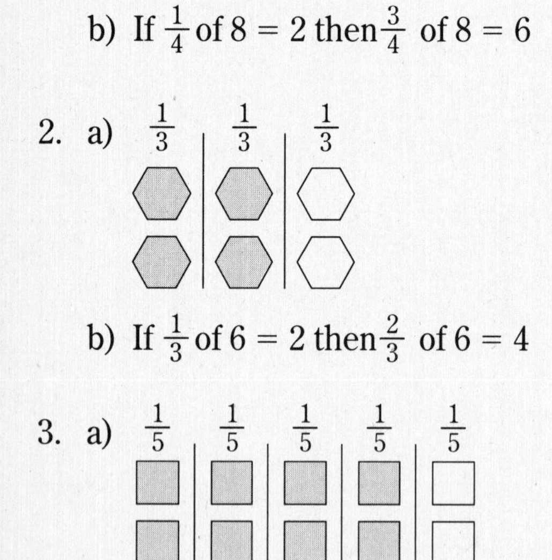

2. a) $\frac{1}{3}$ $\frac{1}{3}$ $\frac{1}{3}$

   b) If $\frac{1}{3}$ of 6 = 2 then $\frac{2}{3}$ of 6 = 4

3. a) $\frac{1}{5}$ $\frac{1}{5}$ $\frac{1}{5}$ $\frac{1}{5}$ $\frac{1}{5}$

   b) If $\frac{1}{5}$ of 10 = 2 then $\frac{4}{5}$ of 10 = 8

Page 6: Apply Your Skills

1. If $\frac{1}{4}$ of 24 = 6 then $\frac{3}{4}$ of 24 = 18
2. If $\frac{1}{5}$ of 15 = 3 then $\frac{3}{5}$ of 15 = 9
3. If $\frac{1}{5}$ of 20 = 4 then $\frac{4}{5}$ of 20 = 16
4. If $\frac{1}{9}$ of 45 = 5 then $\frac{5}{9}$ of 45 = 25
5. If $\frac{1}{7}$ of 63 = 9 then $\frac{6}{7}$ of 63 = 54
6. If $\frac{1}{8}$ of 48 = 6 then $\frac{7}{8}$ of 48 = 42
7. If $\frac{1}{5}$ of 5 = 1 then $\frac{4}{5}$ of 5 = 4
8. If $\frac{1}{10}$ of 100 = 10 then $\frac{7}{10}$ of 100 = 70

Page 7: Multiplication Is Repeated Addition

1. $\frac{1}{2} + \frac{1}{2} + \frac{1}{2} + \frac{1}{2} + \frac{1}{2} = \frac{5}{2} = 2\frac{1}{2}$
2. $\frac{2}{5} + \frac{2}{5} + \frac{2}{5} = \frac{6}{5} = 1\frac{1}{5}$
3. $\frac{3}{4} + \frac{3}{4} + \frac{3}{4} + \frac{3}{4} = \frac{12}{4} = 3$
4. $\frac{1}{2} + \frac{1}{2} + \frac{1}{2} = \frac{3}{2} = 1\frac{1}{2}$

Page 8: Multiplication Models

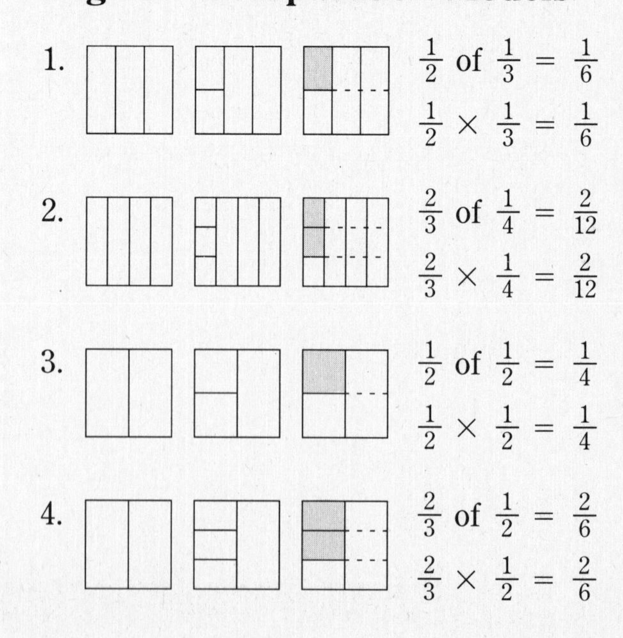

1. $\frac{1}{2}$ of $\frac{1}{3} = \frac{1}{6}$
   $\frac{1}{2} \times \frac{1}{3} = \frac{1}{6}$

2. $\frac{2}{3}$ of $\frac{1}{4} = \frac{2}{12}$
   $\frac{2}{3} \times \frac{1}{4} = \frac{2}{12}$

3. $\frac{1}{2}$ of $\frac{1}{2} = \frac{1}{4}$
   $\frac{1}{2} \times \frac{1}{2} = \frac{1}{4}$

4. $\frac{2}{3}$ of $\frac{1}{2} = \frac{2}{6}$
   $\frac{2}{3} \times \frac{1}{2} = \frac{2}{6}$

CHAPTER 2: MULTIPLICATION

Page 9: Multiplying Fractions and Whole Numbers

1. $1\frac{1}{4}$
2. $1\frac{3}{5}$
3. $2\frac{1}{2}$
4. $1\frac{3}{5}$
5. $5\frac{3}{5}$
6. 3
7. $5\frac{1}{3}$
8. 2
9. $5\frac{2}{5}$
10. $2\frac{2}{5}$
11. $1\frac{3}{5}$
12. $\frac{2}{3}$
13. $5\frac{1}{4}$
14. $1\frac{1}{8}$
15. $1\frac{7}{8}$
16. 18

Page 10: A Fraction Times a Fraction

1. $\frac{1}{24}$
2. $\frac{5}{18}$
3. $\frac{1}{8}$
4. $\frac{1}{6}$
5. $\frac{1}{5}$
6. $\frac{3}{8}$
7. $\frac{1}{2}$
8. $\frac{2}{5}$
9. $\frac{2}{5}$
10. $\frac{1}{20}$
11. $\frac{5}{8}$
12. $\frac{1}{4}$
13. $\frac{2}{21}$
14. $\frac{1}{3}$

Page 11: Simplify the Fractions

1. $\frac{5}{\underset{3}{\cancel{6}}} \times \frac{\overset{2}{\cancel{4}}}{7}$
2. $\frac{\overset{1}{\cancel{5}}}{9} \times \frac{2}{\underset{3}{\cancel{15}}}$
3. $\frac{1}{\underset{1}{\cancel{2}}} \times \frac{\overset{3}{\cancel{6}}}{7}$
4. $\frac{1}{\underset{4}{\cancel{12}}} \times \frac{\overset{1}{\cancel{3}}}{5}$
5. $\frac{\overset{3}{\cancel{12}}}{1} \times \frac{3}{\underset{1}{\cancel{4}}}$
6. $\frac{3}{\underset{2}{\cancel{8}}} \times \frac{\overset{3}{\cancel{12}}}{5}$
7. $\frac{1}{4} \times \frac{\overset{3}{\cancel{9}}}{\underset{5}{\cancel{15}}}$
8. $\frac{\overset{1}{\cancel{4}}}{3} \times \frac{17}{\underset{2}{\cancel{8}}}$
9. $\frac{\overset{1}{\cancel{2}}}{3} \times \frac{7}{\underset{4}{\cancel{8}}}$
10. $\frac{9}{\underset{1}{\cancel{10}}} \times \frac{\overset{1}{\cancel{10}}}{11}$
11. $\frac{4}{\underset{1}{\cancel{5}}} \times \frac{\overset{3}{\cancel{15}}}{17}$
12. $\frac{\overset{2}{\cancel{8}}}{9} \times \frac{1}{\underset{1}{\cancel{4}}}$

Page 12: Simplify First

1. $\frac{\overset{1}{\cancel{4}}}{\underset{3}{\cancel{9}}} \times \frac{\overset{1}{\cancel{3}}}{\underset{2}{\cancel{8}}}$
2. $\frac{\overset{3}{\cancel{18}}}{\underset{1}{\cancel{5}}} \times \frac{\overset{1}{\cancel{5}}}{\underset{1}{\cancel{6}}}$
3. $\frac{\overset{1}{\cancel{5}}}{\underset{6}{\cancel{18}}} \times \frac{\overset{1}{\cancel{3}}}{\underset{4}{\cancel{20}}}$
4. $\frac{\overset{3}{\cancel{6}}}{\underset{5}{\cancel{25}}} \times \frac{\overset{1}{\cancel{5}}}{\underset{4}{\cancel{8}}}$
5. $\frac{\overset{1}{\cancel{2}}}{\underset{1}{\cancel{3}}} \times \frac{\overset{5}{\cancel{15}}}{\underset{8}{\cancel{16}}}$
6. $\frac{\overset{1}{\cancel{4}}}{\underset{1}{\cancel{5}}} \times \frac{\overset{5}{\cancel{25}}}{\underset{8}{\cancel{32}}}$
7. $\frac{\overset{2}{\cancel{14}}}{\underset{3}{\cancel{15}}} \times \frac{\overset{4}{\cancel{20}}}{\underset{3}{\cancel{21}}}$
8. $\frac{\overset{2}{\cancel{8}}}{\underset{3}{\cancel{9}}} \times \frac{\overset{1}{\cancel{3}}}{\underset{1}{\cancel{4}}}$
9. $\frac{\overset{1}{\cancel{32}}}{\underset{5}{\cancel{35}}} \times \frac{\overset{3}{\cancel{21}}}{\underset{1}{\cancel{32}}}$
10. $\frac{\overset{5}{\cancel{35}}}{\underset{3}{\cancel{48}}} \times \frac{\overset{1}{\cancel{16}}}{\underset{3}{\cancel{21}}}$
11. $\frac{\overset{1}{\cancel{11}}}{\underset{3}{\cancel{12}}} \times \frac{\overset{1}{\cancel{4}}}{\underset{3}{\cancel{33}}}$
12. $\frac{\overset{3}{\cancel{24}}}{\underset{5}{\cancel{25}}} \times \frac{\overset{3}{\cancel{15}}}{\underset{4}{\cancel{32}}}$

Page 13: Simplify Before Multiplying

1. $\frac{1}{8}$
2. $\frac{1}{4}$
3. $\frac{4}{7}$
4. $\frac{2}{7}$
5. $12\frac{1}{4}$
6. $\frac{1}{3}$
7. $\frac{2}{3}$
8. $4\frac{4}{5}$
9. $\frac{7}{10}$
10. $\frac{3}{14}$
11. 2
12. $\frac{1}{6}$
13. $\frac{1}{6}$
14. $\frac{1}{10}$
15. $\frac{1}{3}$

Page 14: Practice Your Skills

1. $\frac{5}{14}$
2. $3\frac{1}{3}$
3. $\frac{1}{9}$
4. $1\frac{2}{5}$
5. $\frac{3}{10}$
6. $\frac{1}{3}$
7. $\frac{2}{5}$
8. $\frac{1}{2}$
9. $\frac{1}{9}$
10. 6
11. $2\frac{1}{2}$
12. $\frac{1}{2}$
13. $3\frac{1}{3}$
14. $4\frac{1}{5}$
15. $\frac{3}{25}$
16. $\frac{5}{8}$
17. $3\frac{3}{4}$
18. $\frac{1}{6}$
19. $\frac{1}{3}$
20. $\frac{2}{15}$

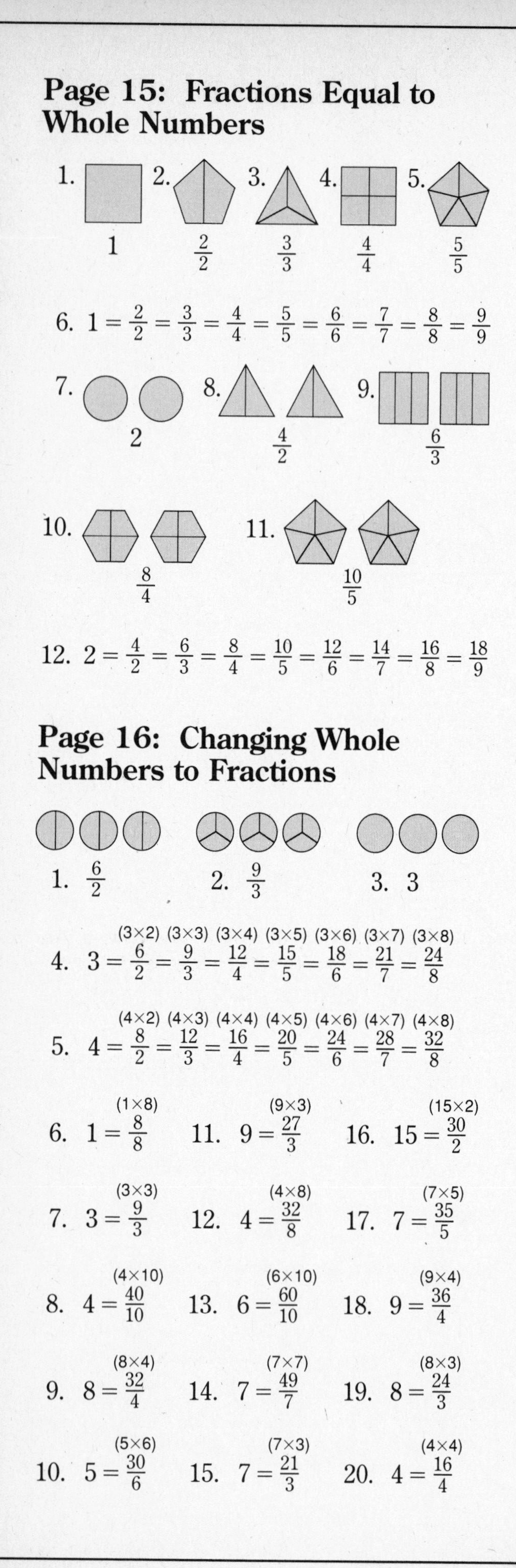

Page 15: Fractions Equal to Whole Numbers

1. 1 2. $\frac{2}{2}$ 3. $\frac{3}{3}$ 4. $\frac{4}{4}$ 5. $\frac{5}{5}$

6. $1 = \frac{2}{2} = \frac{3}{3} = \frac{4}{4} = \frac{5}{5} = \frac{6}{6} = \frac{7}{7} = \frac{8}{8} = \frac{9}{9}$

7. 2 8. $\frac{4}{2}$ 9. $\frac{6}{3}$

10. $\frac{8}{4}$ 11. $\frac{10}{5}$

12. $2 = \frac{4}{2} = \frac{6}{3} = \frac{8}{4} = \frac{10}{5} = \frac{12}{6} = \frac{14}{7} = \frac{16}{8} = \frac{18}{9}$

Page 16: Changing Whole Numbers to Fractions

1. $\frac{6}{2}$ 2. $\frac{9}{3}$ 3. 3

4. $3 = \overset{(3\times2)}{\frac{6}{2}} = \overset{(3\times3)}{\frac{9}{3}} = \overset{(3\times4)}{\frac{12}{4}} = \overset{(3\times5)}{\frac{15}{5}} = \overset{(3\times6)}{\frac{18}{6}} = \overset{(3\times7)}{\frac{21}{7}} = \overset{(3\times8)}{\frac{24}{8}}$

5. $4 = \overset{(4\times2)}{\frac{8}{2}} = \overset{(4\times3)}{\frac{12}{3}} = \overset{(4\times4)}{\frac{16}{4}} = \overset{(4\times5)}{\frac{20}{5}} = \overset{(4\times6)}{\frac{24}{6}} = \overset{(4\times7)}{\frac{28}{7}} = \overset{(4\times8)}{\frac{32}{8}}$

6. $1 = \overset{(1\times8)}{\frac{8}{8}}$ 11. $9 = \overset{(9\times3)}{\frac{27}{3}}$ 16. $15 = \overset{(15\times2)}{\frac{30}{2}}$

7. $3 = \overset{(3\times3)}{\frac{9}{3}}$ 12. $4 = \overset{(4\times8)}{\frac{32}{8}}$ 17. $7 = \overset{(7\times5)}{\frac{35}{5}}$

8. $4 = \overset{(4\times10)}{\frac{40}{10}}$ 13. $6 = \overset{(6\times10)}{\frac{60}{10}}$ 18. $9 = \overset{(9\times4)}{\frac{36}{4}}$

9. $8 = \overset{(8\times4)}{\frac{32}{4}}$ 14. $7 = \overset{(7\times7)}{\frac{49}{7}}$ 19. $8 = \overset{(8\times3)}{\frac{24}{3}}$

10. $5 = \overset{(5\times6)}{\frac{30}{6}}$ 15. $7 = \overset{(7\times3)}{\frac{21}{3}}$ 20. $4 = \overset{(4\times4)}{\frac{16}{4}}$

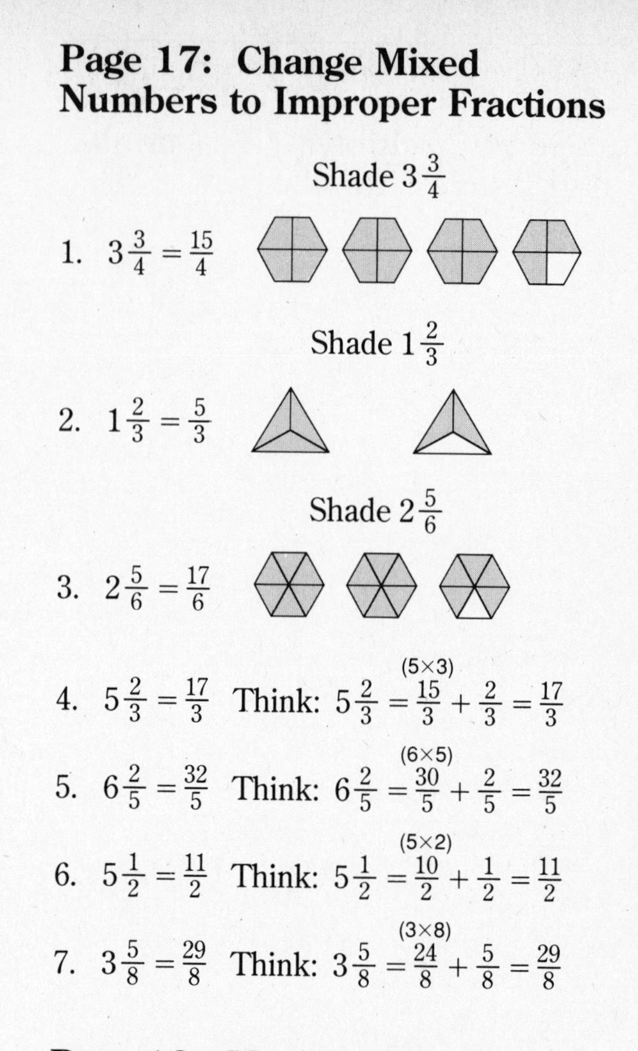

Page 17: Change Mixed Numbers to Improper Fractions

Shade $3\frac{3}{4}$

1. $3\frac{3}{4} = \frac{15}{4}$

Shade $1\frac{2}{3}$

2. $1\frac{2}{3} = \frac{5}{3}$

Shade $2\frac{5}{6}$

3. $2\frac{5}{6} = \frac{17}{6}$

4. $5\frac{2}{3} = \frac{17}{3}$ Think: $5\frac{2}{3} = \overset{(5\times3)}{\frac{15}{3}} + \frac{2}{3} = \frac{17}{3}$

5. $6\frac{2}{5} = \frac{32}{5}$ Think: $6\frac{2}{5} = \overset{(6\times5)}{\frac{30}{5}} + \frac{2}{5} = \frac{32}{5}$

6. $5\frac{1}{2} = \frac{11}{2}$ Think: $5\frac{1}{2} = \overset{(5\times2)}{\frac{10}{2}} + \frac{1}{2} = \frac{11}{2}$

7. $3\frac{5}{8} = \frac{29}{8}$ Think: $3\frac{5}{8} = \overset{(3\times8)}{\frac{24}{8}} + \frac{5}{8} = \frac{29}{8}$

Page 18: More Practice Changing to Improper Fractions

1. $\frac{5}{2}$
2. $\frac{41}{6}$
3. $\frac{14}{9}$
4. $\frac{9}{4}$
5. $\frac{51}{7}$
6. $\frac{29}{3}$
7. $\frac{28}{3}$
8. $\frac{21}{8}$
9. $\frac{23}{4}$
10. $\frac{47}{6}$
11. $\frac{26}{5}$
12. $\frac{23}{6}$
13. $\frac{11}{3}$
14. $\frac{25}{2}$
15. $\frac{39}{4}$
16. $\frac{47}{3}$
17. $\frac{33}{8}$
18. $\frac{42}{5}$

Page 19: Rename the Mixed Number

1. $\frac{11}{21}$
2. $2\frac{4}{5}$
3. $\frac{33}{50}$
4. $3\frac{1}{5}$
5. 2
6. $\frac{1}{2}$

Page 19: Rename the Mixed Number (continued)

7. $1\frac{1}{2}$
8. $2\frac{1}{6}$
9. $2\frac{1}{8}$
10. $1\frac{1}{35}$

Page 20: Multiplying Mixed Numbers

1. $12\frac{2}{3}$
2. $3\frac{5}{8}$
3. 8
4. $9\frac{1}{5}$
5. 24
6. $7\frac{2}{5}$
7. $7\frac{2}{9}$
8. $6\frac{4}{7}$

Page 21: Master the Skills

1. $1\frac{2}{5}$
2. $\frac{3}{5}$
3. $2\frac{1}{2}$
4. 10
5. $\frac{2}{15}$
6. $\frac{1}{4}$
7. $\frac{3}{5}$
8. $1\frac{11}{12}$
9. $1\frac{1}{14}$
10. $\frac{11}{21}$
11. $6\frac{1}{4}$
12. 14
13. $\frac{2}{3}$
14. $1\frac{5}{8}$
15. $6\frac{2}{3}$
16. $5\frac{1}{4}$

Page 22: Multiplication Review

1. \$5.00
2. 5
3. a)

$\frac{1}{4}$ $\frac{1}{4}$ $\frac{1}{4}$ $\frac{1}{4}$

b) If $\frac{1}{4}$ of 8 = 2, then $\frac{3}{4}$ of 8 = 6.

4. If $\frac{1}{4}$ of 36 = 9, then $\frac{3}{4}$ of 36 = 27.
5. $\frac{6}{5} = 1\frac{1}{5}$
6. $\frac{2}{3}$ of $\frac{1}{3} = \frac{2}{3} \times \frac{1}{3}$
7. $2\frac{2}{5}$

Page 22: Multiplication Review (continued)

8. $\frac{7}{24}$
9. $\frac{1}{\underset{1}{\cancel{5}}} \times \frac{\overset{3}{\cancel{15}}}{17} = \frac{3}{17}$
10. $\frac{\overset{2}{\cancel{8}}}{\underset{3}{\cancel{9}}} \times \frac{\overset{1}{\cancel{3}}}{\underset{1}{\cancel{4}}} = \frac{2}{3}$
11. $\frac{\overset{1}{\cancel{5}}}{\underset{9}{\cancel{18}}} \times \frac{\overset{1}{\cancel{2}}}{\underset{1}{\cancel{5}}} = \frac{1}{9}$
12. $\frac{2}{\underset{1}{\cancel{3}}} \times \frac{\overset{2}{\cancel{6}}}{1} = \frac{4}{1} = 4$
13. $\frac{12}{6}$
14. $\frac{28}{4}$
15. $\frac{23}{5}$
16. $\frac{49}{8}$
17. $2\frac{2}{5} \times \frac{5}{6} = \frac{\overset{2}{\cancel{12}}}{\underset{1}{\cancel{5}}} \times \frac{\overset{1}{\cancel{5}}}{\underset{1}{\cancel{6}}} = \frac{2}{1} = 2$
18. $5\frac{1}{3} \times 5\frac{1}{4} = \frac{\overset{4}{\cancel{16}}}{\underset{1}{\cancel{3}}} \times \frac{\overset{7}{\cancel{21}}}{\underset{1}{\cancel{4}}} = \frac{28}{1} = 28$

CHAPTER 3: MULTIPLICATION PROBLEM SOLVING

Page 23: Find a Fraction of an Amount

1. $21\frac{1}{3}$
2. 3
3. \$22.00
4. 3

Page 24: Does the Answer Make Sense?

1. $\frac{1}{2} \times 5 = 2\frac{1}{2}$; $2\frac{1}{2}$
2. $2\frac{2}{3} \times \$3 = \$8$; \$8
3. $24 \times \frac{3}{4} = 18$; 18

Page 24: Does the Answer Make Sense? (continued)

4. $6\frac{1}{2} \times 10 = 65; 65$

5. $\frac{1}{5} \times \$145 = \$29; \$29$

6. $3\frac{3}{4} \times 6 = 22\frac{1}{2}; 22\frac{1}{2}$

Page 25: Number Sentences

1. $\frac{1}{2} \times 4 = 2; 2$

2. $\frac{3}{4} \times 8 = 6; 6$

3. $\frac{3}{4} \times 4 = 3; 3$

4. $16\frac{1}{2} \times \frac{1}{3} = 5\frac{1}{2}; 5\frac{1}{2}$

5. $\frac{5}{8} \times 48 = 30; 30$

6. $\frac{3}{4} \times 28 = 21; 21$

7. $24 \times \frac{1}{3} = \$8.00; \$8.00$

8. $6\frac{1}{2} \times 6 = 39; 39$

CHAPTER 4: UNDERSTANDING DIVISION OF FRACTIONS

Page 26: Divide Whole Numbers by Fractions

1. a) 10
   b) 10
2. a) 12
   b) 12
3. a) 9
   b) 9
4. a) 8
   b) 8
5. a) 20
   b) 20
6. a) 24
   b) 24

Page 27: Think About Fraction Division

1. $2 \div \frac{1}{4} = 8$

2. $2 \times 4 = 8$ $\quad \frac{4}{4}$ or 1 $\quad \frac{4}{4}$ or 1

Page 27: Think About Fraction Division (continued)

3. $3 \div \frac{1}{2} = 6$

4. $3 \times 2 = 6$ $\quad \frac{2}{2}$ or 1 $\quad \frac{2}{2}$ or 1 $\quad \frac{2}{2}$ or 1

5. 4
6. 4
7. 6
8. 12
9. 10
10. 20
11. 8
12. 9
13. 15
14. 12
15. 16
16. 8

Page 28: Reciprocals

	Reciprocal
1.	$\frac{5}{3}$
2.	$\frac{8}{3}$
3.	$\frac{2}{8}$
4.	$\frac{4}{3}$
5.	$\frac{3}{5}$

	Number	Reciprocal
6.	$5 = \frac{5}{1}$	$\frac{1}{5}$
7.	$7 = \frac{7}{1}$	$\frac{1}{7}$
8.	$3 = \frac{3}{1}$	$\frac{1}{3}$
9.	$4 = \frac{4}{1}$	$\frac{1}{4}$
10.	$15 = \frac{15}{1}$	$\frac{1}{15}$
11.	$3\frac{1}{2} = \frac{7}{2}$	$\frac{2}{7}$
12.	$2\frac{3}{4} = \frac{11}{4}$	$\frac{4}{11}$
13.	$4\frac{1}{5} = \frac{21}{5}$	$\frac{5}{21}$
14.	$3\frac{5}{6} = \frac{23}{6}$	$\frac{6}{23}$
15.	$1\frac{3}{5} = \frac{8}{5}$	$\frac{5}{8}$

Page 29: Dividing with Fractions

1. a) 6
   b) 6
   c) $\frac{8}{1}$
2. a) 4
   b) 4
   c) $\frac{6}{1}$
3. a) 2
   b) 2
   c) $\frac{4}{1}$
4. a) 4
   b) 4
   c) $\frac{8}{1}$

CHAPTER 5: DIVISION

Page 30: Multiply by the Reciprocal

A. $\frac{8}{9}$

B. $\frac{16}{35}$

1. $\frac{4}{5}$
2. $\frac{3}{8}$
3. $\frac{2}{3} \times \frac{8}{7} = \frac{16}{21}$
4. $\frac{1}{8} \times \frac{3}{2} = \frac{3}{16}$
5. $\frac{4}{5} \times \frac{8}{7} = \frac{32}{35}$
6. $\frac{1}{5} \times \frac{3}{1} = \frac{3}{5}$
7. $\frac{8}{35}$
8. $\frac{24}{35}$
9. $\frac{1}{2} \times \frac{5}{3} = \frac{5}{6}$
10. $\frac{2}{5} \times \frac{4}{3} = \frac{8}{15}$
11. $\frac{3}{4} \times \frac{5}{4} = \frac{15}{16}$
12. $\frac{1}{6} \times \frac{5}{1} = \frac{5}{6}$

Page 31: A Fraction Divided by a Fraction

1. $\frac{1}{5}$
2. $1\frac{1}{4}$
3. $\frac{3}{8}$
4. $1\frac{1}{3}$
5. $\frac{4}{5}$
6. $2\frac{2}{5}$
7. $\frac{3}{4}$
8. $1\frac{7}{8}$
9. $\frac{1}{2}$
10. $3\frac{1}{3}$
11. $1\frac{1}{15}$
12. 3

Page 32: Dividing by a Fraction

1. a) 6
   b) 6
2. a) 5
   b) 5
3. a) 3
   b) 3
4. a) 3
   b) 3
5. a) 2
   b) 2

Page 33: Think About Dividing by Fractions

1. a) 3
   b) 3
2. a) 5
   b) 5
3. a) 7
   b) 7
4. a) 9
   b) 9
5. a) 2
   b) 2

Page 34: Using Drawings

A. 5
B. 5

1. a) 3
   b) 3
2. a) 7
   b) 7
3. a) 10
   b) 10
4. a) 14
   b) 14

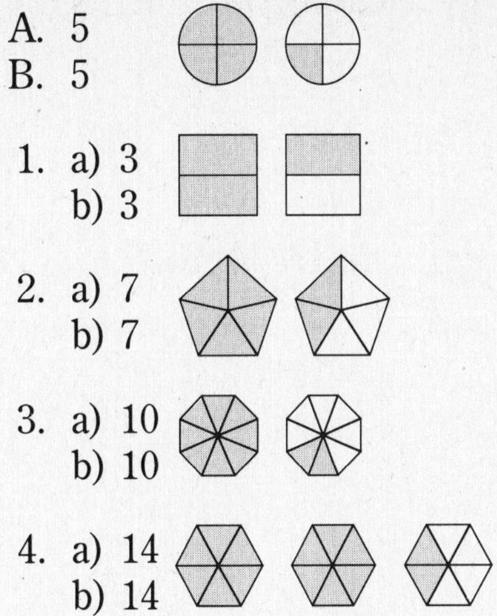

Page 35: Dividing a Mixed Number by a Fraction

1. $3\frac{3}{4}$
2. $3\frac{5}{9}$
3. $7\frac{1}{2}$
4. $5\frac{1}{4}$
5. $2\frac{2}{5}$
6. 30
7. $13\frac{1}{3}$
8. $5\frac{3}{5}$
9. 5
10. 2
11. 25
12. $12\frac{3}{4}$

Page 36: Mixed Practice

1. 16
2. $\frac{6}{7}$
3. 14
4. $\frac{1}{3}$
5. $17\frac{1}{2}$
6. 6
7. 25
8. 7
9. 10
10. $1\frac{1}{2}$
11. 4
12. 10
13. 4
14. $9\frac{1}{3}$
15. 3
16. $1\frac{1}{2}$

Page 37: Divide Whole Numbers by Fractions

1. 27
2. 12
3. $17\frac{1}{2}$
4. 18
5. 21
6. 20
7. 6
8. 22
9. 12

Page 37: Divide Whole Numbers by Fractions (continued)

10. 49
11. 15
12. $7\frac{1}{5}$

Page 38: Divide by Whole Numbers

1. $\frac{7}{12}$
2. $\frac{3}{8}$
3. $\frac{3}{25}$
4. $\frac{2}{5}$
5. $1\frac{2}{3}$
6. $2\frac{1}{5}$
7. $\frac{6}{7}$
8. $\frac{7}{8}$
9. $2\frac{1}{2}$
10. $\frac{5}{9}$
11. $1\frac{1}{3}$
12. $1\frac{3}{4}$

Page 39: Divide by Mixed Numbers

1. $\frac{1}{4}$
2. $\frac{1}{5}$
3. $\frac{1}{16}$
4. $\frac{6}{25}$
5. $\frac{3}{20}$
6. $\frac{7}{30}$
7. $\frac{2}{3}$
8. $\frac{1}{6}$
9. $\frac{18}{49}$
10. $\frac{3}{37}$
11. $\frac{8}{27}$
12. $\frac{2}{11}$

Page 40: Divide Two Mixed Numbers

1. $1\frac{7}{8}$
2. 2
3. $\frac{3}{5}$
4. 3
5. $\frac{13}{16}$
6. $\frac{13}{19}$
7. $1\frac{2}{7}$
8. $2\frac{8}{11}$
9. 7
10. $3\frac{3}{4}$
11. $2\frac{2}{5}$
12. $5\frac{5}{6}$

Page 41: Division Practice

1. 6
2. $\frac{15}{22}$
3. $\frac{8}{27}$
4. $3\frac{3}{4}$
5. $2\frac{7}{9}$
6. 6
7. $\frac{1}{10}$
8. $1\frac{1}{3}$
9. $1\frac{7}{15}$
10. $2\frac{1}{2}$

Page 42: Division Review

1. 8
2. 15
3. $\frac{5}{19}$
4. $1\frac{1}{2}$
5. $2\frac{2}{3}$
6. 5
7. $4\frac{2}{7}$
8. 6
9. 55
10. $\frac{5}{8}$
11. $\frac{3}{5}$
12. 2

Page 43: Use All Operations

1. $2\frac{2}{5}$
2. $\frac{11}{16}$
3. $3\frac{4}{5}$
4. $\frac{5}{8}$
5. $7\frac{2}{3}$
6. 9
7. $14\frac{5}{12}$
8. $4\frac{3}{4}$
9. $1\frac{1}{15}$
10. 15
11. $8\frac{1}{4}$
12. $\frac{3}{14}$

Page 44: Putting It All Together

1. $\frac{1}{4} < 3\frac{1}{3}$
2. $3\frac{1}{3} > 3\frac{1}{11}$
3. $5\frac{5}{18} > 4\frac{7}{10}$
4. $4\frac{1}{2} > 3\frac{1}{23}$
5. $5\frac{7}{8} > 5\frac{3}{8}$
6. $\frac{1}{8} < \frac{3}{8}$
7. $5\frac{7}{8} > 4\frac{3}{8}$
8. $\frac{13}{16} < 1\frac{1}{2}$

CHAPTER 6: MIXED PROBLEM SOLVING

Page 45: Does the Answer Make Sense?

1. $4 \div \frac{1}{4} = 16$; 16
2. $1\frac{1}{4} \times 3\frac{3}{4} = 4\frac{11}{16}$; $4\frac{11}{16}$
3. $12 \div \frac{3}{4} = 16$; 16
4. $1\frac{1}{4} \times 7\frac{1}{2} = 9\frac{3}{8}$; $9\frac{3}{8}$
5. $15 \div \frac{3}{4} = 20$; 20
6. $6 \div \frac{2}{5} = 15$; 15

Page 46: Decide to Multiply or Divide

1. $\frac{1}{2} \times 5 = 2\frac{1}{2}; 2\frac{1}{2}$
2. $35 \times \frac{1}{5} = \$7.00; \$7.00$
3. $5\frac{1}{2} \div 2 = 2\frac{3}{4}; 2\frac{3}{4}$
4. $5 \div \frac{1}{2} = 10; 10$

Page 47: Mixed Multiplication and Division

1. $3\frac{1}{2} \div \frac{1}{4} = 14; 14$
2. $\frac{1}{5} \times 7\frac{1}{2} = 1\frac{1}{2}; 1\frac{1}{2}$
3. $3\frac{1}{3} \times 2\frac{1}{4} = 7\frac{1}{2}; 7\frac{1}{2}$
4. $1\frac{1}{2} \div 2 = \frac{3}{4}; \frac{3}{4}$
5. $10 \div 2\frac{1}{2} = 4; 4$
6. $5 \times 8\frac{1}{2} = 42\frac{1}{2}; 42\frac{1}{2}$
7. $25 \div 2\frac{1}{2} = 10; 10$
8. $\frac{3}{4} \times 2 = 1\frac{1}{2}; 1\frac{1}{2}$

Page 48: Think It Through

Answers should be similar to these.

1. a) How many cups of flour and milk are used in all?

   b) How many more cups of flour were used?

2. a) How much was marked off?

   b) What was the sale price of the sweater?

3. a) How many hours did Bob work in two $8\frac{1}{2}$-hour shifts?

   b) If Bob worked three $8\frac{1}{2}$-hour shifts, how many hours did he work?

Page 48: Think It Through (continued)

4. a) About how many hours did Bonita travel each hour?

   b) If she travels at the same speed, about how many miles would Bonita travel in 5 hours?

5. a) 5 centimeters are about how many inches?

   b) 3 centimeters are about how many inches?

6. a) How much does each box weigh?

   b) How much do 5 boxes weigh?

Page 49: Write a Question

Sample Questions

1. How much money was marked off?

   $36 \times \frac{1}{4} = \$9.00$

2. How many miles did he walk in all?

   $14 \times 1\frac{1}{2} = 21$ miles

3. How many $1\frac{1}{2}$-foot pieces will he have?

   $6 \div 1\frac{1}{2} = 4$ pieces

4. What is the total weight of 12 sacks?

   $\frac{1}{2} \times 12 = 6$ pounds

5. About how many centimeters are in 20 inches?

   $20 \div \frac{2}{5} = 50$ centimeters

6. How much does Lynn save each week?

   $125 \times \frac{1}{5} = \$25.00$

Page 50: Apply the Operations

1. D
2. B
3. A
4. C
5. $\frac{1}{2} \times 1\frac{1}{4} = \frac{5}{8}; \frac{5}{8}$
6. $15 \div 3\frac{3}{4} = 4; 4$
7. $1\frac{1}{2} + 3\frac{3}{4} = 5\frac{1}{4}; 5\frac{1}{4}$
8. $8\frac{1}{2} - 1\frac{1}{8} = 7\frac{3}{8}; 7\frac{3}{8}$

Page 51: Choose the Operation

1. addition
2. subtraction
3. multiplication
4. multiplication
5. division
6. addition
7. division
8. multiplication
9. subtraction
10. addition

Page 52: Mixed Problem Solving

1. a) $\frac{5}{8}$ of the strawberries were picked
   b) $\frac{3}{8}$ left to pick
2. $3\frac{1}{2} - 1\frac{1}{4} = 2\frac{1}{4}; 2\frac{1}{4}$
3. $3\frac{1}{4} + 2\frac{1}{3} = 5\frac{7}{12}; 5\frac{7}{12}$
4. $\frac{2}{3} \times \frac{1}{2} = \frac{1}{3}; \frac{1}{3}$
5. $\frac{1}{10} + \frac{1}{5} = \frac{3}{10}; \frac{3}{10}$
6. $4 \div \frac{1}{2} = 8; 8$

Page 53: Throw Away Extra Information

1. a) Facts not needed: $3.98; 7 people
   b) You don't need the number of people or the price to figure out number of pounds.
   c) $1\frac{1}{4}$ pounds left
2. a) Facts not needed: 225 miles; 485 miles
   b) You don't need the number of miles to figure out number of hours.

Page 53: Throw Away Extra Information (continued)

   c) $12\frac{5}{6}$ hours traveled
3. a) Facts not needed: $15.98; 4 people; 12 friends
   b) You don't need the price and number of people to figure out number of pounds.
   c) $3\frac{3}{4}$ pounds used

Page 54: Two-Step Story Problems

A. $27 \times \frac{1}{3} = \$9$

B. $\$27 - \$9 = \$18$

1. a) $10 \div 2\frac{1}{2} = 4$ boards
   b) $20 \div 4 = 5$ boards 10 feet long
2. a) $180 \div 60 = 3$ batches
   b) $1\frac{1}{4} \times 3 = 3\frac{3}{4}$ cups
3. a) $40 \times \frac{4}{5} = 32$ questions
   b) $40 \times \frac{7}{8} = 35$ questions
   c) yes
4. a) $60 \times \frac{1}{5} = 12$ cookies
   b) $60 - 12 = 48$ cookies

Page 55: More Two-Step Problems

A. $6\frac{3}{4}$ gallons

B. $2\frac{1}{4}$ gallons

1. 4 students have the flu
2. $25.00
3. 20
4. 4

Page 56: Multi-Step Word Problems

Q1: $2\frac{1}{4}$ hours

Q2: $2\frac{3}{4}$ hours

Q3: $\frac{1}{2}$ hour

Answers should be similar to these.

1. Q1: How much does he spend on rent? $145

   Q2: How much does he spend on food? $87

   Q3: After subtracting food and rent, how much does he have left? $203

2. Q1: How long is the first shift? 6 hours

   Q2: How long is the second shift? 8 hours

   Q3: How many hours of the day are left for the third shift? 10 hours

3. Q1: How long did Walter take altogether? $17\frac{1}{4}$ hours

   Q2: How long did Caren take altogether? 18 hours

   Q3: How much longer did Caren take? $\frac{3}{4}$ hour

4. Q1: How much flour did she use? 4 lbs.

   Q2: How much flour is left over? $2\frac{1}{2}$ lbs.

   Q3: How many times can she fill the bread recipe with the remaining flour? 2 times

CHAPTER 7: LIFE-SKILLS MATH

Page 57: Decrease a Recipe

1. a) $\frac{3}{4}$ b) $1\frac{1}{2}$ c) $\frac{1}{8}$ d) $\frac{5}{8}$ e) $\frac{3}{8}$ f) $\frac{1}{6}$ g) 1
2. 6, or $\frac{1}{2}$ dozen
3. a) 12 b) 6 batches

Page 58: Increase a Recipe

1. a) 5 b) $2\frac{2}{3}$ c) 2 d) 1 e) $1\frac{1}{2}$ f) 1 g) 16 h) 4
2. 2 batches
3. a) 4 b) $1\frac{1}{2}$

Page 59: At the Store

Cost Per Pound	Cost
1. a) $1.04	b) $ 3.64
2. a) .92	b) 2.53
3. a) .51	b) .68
4. a) 1.25	b) 8.75
5. a) 1.44	b) 1.62
Total Cost	6. $17.22

7. $6.40
8. $2.02
9. $3.62
10. $.86

Page 60: Common Discounts

1. $142.44
2. $26.63
3. $7.49
4. a) $8.00 b) $4.13 c) $9.00 d) $1.33 e) $49.00 f) $299.50

CHAPTER 1: MEANING OF RATIO

Page 1: What Is Ratio?

1. a) 5 to 3 b) 5 : 3 c) $\frac{5}{3}$
2. a) 3 to 8 b) 3 : 8 c) $\frac{3}{8}$
3. a) 5 to 8 b) 5 : 8 c) $\frac{5}{8}$

Page 2: Write the Ratios

1. a) 2 to 3 b) 2 : 3 c) $\frac{2}{3}$
2. a) 3 to 2 b) 3 : 2 c) $\frac{3}{2}$
3. a) 2 to 5 b) 2 : 5 c) $\frac{2}{5}$
4. a) 3 to 5 b) 3 : 5 c) $\frac{3}{5}$
5. a) 5 to 2 b) 5 : 2 c) $\frac{5}{2}$

Page 3: Ratios as Fractions

1. $\frac{1}{5}$ 3. $\frac{3}{4}$ 5. $\frac{5}{7}$
2. $\frac{2}{6}$ 4. $\frac{1}{3}$ 6. $\frac{2}{5}$

Page 4: Compare the Shapes

1. $4:7 = \frac{4}{7}$
2. $7:4 = \frac{7}{4}$
3. $4:11 = \frac{4}{11}$
4. $7:11 = \frac{7}{11}$
5. $11:4 = \frac{11}{4}$
6. $11:7 = \frac{11}{7}$
7. $5:7 = \frac{5}{7}$
8. $7:5 = \frac{7}{5}$

Page 5: Draw the Ratios

1. $\frac{2}{6}$

Page 5: Draw the Ratios (continued)

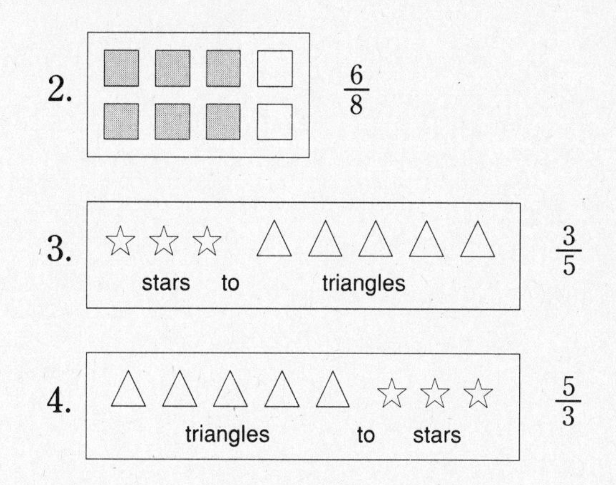

2. $\frac{6}{8}$
3. $\frac{3}{5}$
4. $\frac{5}{3}$

Page 6: Write the Ratios as Fractions

1. $\frac{17}{30}$ 3. $\frac{10}{2}$ 5. $\frac{5}{12}$ 7. $\frac{3}{10}$
2. $\frac{2}{3}$ 4. $\frac{384}{7}$ 6. $\frac{100}{.89}$ 8. $\frac{6}{10}$

Page 7: Simplify the Ratios

1. $\frac{5}{20} = \frac{1}{4}$
2. $\frac{12}{15} = \frac{4}{5}$
3. $\frac{15}{30} = \frac{1}{2}$
4. $\frac{4}{12} = \frac{1}{3}$
5. $\frac{8}{64} = \frac{1}{8}$
6. $\frac{6}{8} = \frac{3}{4}$
7. $\frac{8}{24} = \frac{1}{3}$
8. $\frac{16}{18} = \frac{8}{9}$
9. $\frac{9}{36} = \frac{1}{4}$
10. $\frac{21}{24} = \frac{7}{8}$
11. $\frac{7}{49} = \frac{1}{7}$
12. $\frac{6}{18} = \frac{1}{3}$
13. $\frac{28}{35} = \frac{4}{5}$
14. $\frac{10}{25} = \frac{2}{5}$
15. $\frac{11}{22} = \frac{1}{2}$
16. $\frac{5}{20} = \frac{1}{4}$

Page 8: Denominator of 1

1. $\frac{8}{2} = \frac{4}{1}$
2. $\frac{9}{3} = \frac{3}{1}$
3. $\frac{15}{5} = \frac{3}{1}$
4. $\frac{36}{6} = \frac{6}{1}$
5. $\frac{50}{10} = \frac{5}{1}$
6. $\frac{24}{4} = \frac{6}{1}$

Page 8: Denominator of 1 (continued)

7. $\frac{60}{10} = \frac{6}{1}$

8. $\frac{56}{8} = \frac{7}{1}$

9. $\frac{12}{4} = \frac{3}{1}$

10. $\frac{63}{9} = \frac{7}{1}$

11. $\frac{70}{7} = \frac{10}{1}$

12. $\frac{64}{8} = \frac{8}{1}$

13. $\frac{49}{7} = \frac{7}{1}$

14. $\frac{100}{5} = \frac{20}{1}$

Page 9: Equal Ratios Are Equal Fractions

1. $\frac{1}{3} = \frac{2}{6} = \frac{3}{9} = \frac{4}{12} = \frac{5}{15} = \frac{6}{18} = \frac{7}{21} = \frac{8}{24} = \frac{9}{27}$

$\frac{1}{3} \times \frac{2}{2} = \frac{2}{6}$

2. $\frac{5}{1} = \frac{10}{2} = \frac{15}{3} = \frac{20}{4} = \frac{25}{5} = \frac{30}{6} = \frac{35}{7} = \frac{40}{8} = \frac{45}{9}$

3. $\frac{3}{1} = \frac{6}{2} = \frac{9}{3} = \frac{12}{4} = \frac{15}{5} = \frac{18}{6} = \frac{21}{7} = \frac{24}{8} = \frac{27}{9}$

$\frac{3}{1} \times \frac{3}{3} = \frac{9}{3}$

4. $\frac{1}{5} = \frac{2}{10} = \frac{3}{15} = \frac{4}{20} = \frac{5}{25} = \frac{6}{30} = \frac{7}{35} = \frac{8}{40} = \frac{9}{45}$

Page 10: Equivalent Ratios

1. $\frac{1}{2} = \frac{4}{8}$

2. $\frac{5}{6} = \frac{15}{18}$

3. $\frac{7}{8} = \frac{14}{16}$

4. $\frac{1}{5} = \frac{9}{45}$

5. $\frac{6}{7} = \frac{18}{21}$

6. $\frac{3}{4} = \frac{15}{20}$

7. $\frac{2}{9} = \frac{8}{36}$

8. $\frac{4}{5} = \frac{20}{25}$

9. $\frac{5}{20} = \frac{1}{4}$

10. $\frac{18}{45} = \frac{2}{5}$

11. $\frac{2}{14} = \frac{1}{7}$

12. $\frac{21}{49} = \frac{3}{7}$

13. $\frac{9}{24} = \frac{3}{8}$

14. $\frac{5}{6} = \frac{10}{12}$

15. $\frac{3}{4} = \frac{27}{36}$

16. $\frac{5}{3} = \frac{10}{6}$

17. $\frac{5}{4} = \frac{20}{16}$

18. $\frac{40}{64} = \frac{5}{8}$

19. $\frac{5}{7} = \frac{30}{42}$

20. $\frac{3}{7} = \frac{21}{49}$

Page 10: Equivalent Ratios (continued)

21. $\frac{10}{25} = \frac{2}{5}$

22. $\frac{1}{8} = \frac{5}{40}$

23. $\frac{3}{27} = \frac{1}{9}$

24. $\frac{3}{2} = \frac{6}{4}$

CHAPTER 2: RATIO APPLICATIONS

Page 11: Find a Pattern

1.

Cans	1	2	3	4	5	6	7	8
Tennis Balls	3	6	9	12	15	18	21	24

2.

Packs of Gum	1	3	5	7	9	11	13	15
Sticks of Gum	5	15	25	35	45	55	65	75

$\frac{1 \times 9}{5 \times 9} = \frac{9}{45}$

3.

Touchdowns	1	7	3	2	9	4	6	5
Points	6	42	18	12	54	24	36	30

$\frac{1 \times 2}{6 \times 2} = \frac{2}{12}$

4.

Dollars	1	3	7	2	10	6	9	4
Quarters	4	12	28	8	40	24	36	16

$\frac{1 \times 3}{4 \times 3} = \frac{3}{12}$

Page 12: Fill in the Table

1.

Hours Worked	1	2	3	8	40
Paid	$9	$18	$27	$72	$360

$\frac{1 \times 3}{9 \times 3} = \frac{3}{27}$

2.

Hours Worked	1	2	3	4	5
Paid	$8.75	$17.50	$26.25	$35	$43.75

$\frac{1 \times 4}{8.75 \times 4} = \frac{4}{35}$

Page 12: Fill in the Table (continued)

3.

Hours Worked	1	8	10	40	80
Paid	$6.25	$50	$62.50	$250	$500

4.

Hours Worked	1	4	8	20	40
Paid	$12.95	$51.80	$103.60	$259	$518

Page 13: From Words to Ratios

1. $\frac{8}{2} = \frac{4}{1}$
2. $\frac{3}{12} = \frac{1}{4}$
3. $\frac{24}{3} = \frac{8}{1}$
4. $\frac{4}{36} = \frac{1}{9}$
5. $\frac{100}{5} = \frac{20}{1}$
6. $\frac{6}{10} = \frac{3}{5}$
7. $\frac{5}{35} = \frac{1}{7}$
8. $\frac{27}{3} = \frac{9}{1}$

Page 14: Unit Rates

1. $\frac{50}{1}$ 50 miles (per) hour
2. $\frac{3}{1}$ 3 tennis balls (per) can
3. $\frac{9}{1}$ 9 revolutions (each) minute
4. $\frac{50}{1}$ 50 tablets in (one) bottle
5. $\frac{25}{1}$ 25 miles (per) gallon

Page 15: Write as a Ratio in Fraction Form

1. $\frac{46}{1}$ (per)
2. $\frac{48}{5}$
3. $\frac{7}{20}$
4. $\frac{11}{9}$
5. $\frac{1}{5}$
6. $\frac{7}{1}$ (per)
7. $\frac{1}{28}$
8. $\frac{8}{9.85}$
9. $\frac{32}{1}$ (per)
10. $\frac{9}{11}$
11. $\frac{25}{1}$ (each)
12. $\frac{55}{1}$ (per)
13. $\frac{5}{16}$
14. $\frac{36}{1}$ (per)
15. $\frac{2}{1}$ (per)
16. $\frac{129}{8}$
17. $\frac{9}{1}$ (per)
18. $\frac{5}{1}$ (per)
19. $\frac{17}{1}$ (per)
20. $\frac{25}{3}$

Page 16: Writing Unit Rates

1. $\frac{85}{5} = \frac{17}{1}$ 17 miles per gallon
2. $\frac{165}{3} = \frac{55}{1}$ 55 miles per hour
3. $\frac{64}{4} = \frac{16}{1}$ 16 ounces per can
4. $\frac{56}{7} = \frac{8}{1}$ 8 apples per box
5. $\frac{63}{9} = \frac{7}{1}$ 7 ounces per cup
6. $\frac{21}{3} = \frac{7}{1}$ $7 per ticket
7. $\frac{360}{8} = \frac{45}{1}$ 45 miles per hour
8. $\frac{120}{5} = \frac{24}{1}$ 24 miles per gallon

Page 17: Comparing Unit Rates

1. $\frac{133}{7} = \frac{19}{1}$ $\frac{108}{6} = \frac{18}{1}$

19 miles per hour > 18 miles per hour

---

2. $\frac{72}{3} = \frac{24}{1}$ $\frac{120}{5} = \frac{24}{1}$

24 miles per gallon = 24 miles per gallon

---

3. $\frac{60}{4} = \frac{15}{1}$ $\frac{90}{5} = \frac{18}{1}$

$15 for one ticket < $18 for one ticket

---

4. $\frac{425}{5} = \frac{85}{1}$ $\frac{294}{3} = \frac{98}{1}$

85 meters per hour < 98 meters per hour

Page 18: Find the Rates

1. a) $24.00
   b) $64.00
2. a) 360 miles
   b) 180 miles
3. a) 345 miles
   b) 506 miles
4. a) $4.17
   b) $6.95
5. a) $24.75
   b) $57.75
6. a) $20.00
   b) $45.00

Page 19: Find the Cost

1. a) $.99
   b) $1.98
   c) $2.97
   d) $3.96

Page 19: Find the Cost (continued)

2. a) $1.45
   b) $5.80
   c) $8.70
   d) $14.50

3. a) $1.19
   b) $2.38
   c) $13.09
   d) $17.85

4. a) $.78
   b) $1.56
   c) $3.12
   d) $3.90

5. a) $2.25
   b) $4.50
   c) $11.25
   d) $15.75

6. a) $4.25
   b) $12.75
   c) $17.00
   d) $25.50

Page 20: Ratios as Rates

1. a) $24.00
   b) $36.00
   c) $48.00

2. a) $10.00
   b) $15.00
   c) $20.00

3. a) $1.84
   b) $2.76
   c) $3.68

4. a) $9.00
   b) $13.50
   c) $18.00

5. a) $24.00
   b) $40.00
   c) $56.00

6. a) $4.05
   b) $6.75
   c) $10.80

7. a) $12.75
   b) $21.25
   c) $38.25

8. a) $4.50
   b) $7.50
   c) $12.00

Page 21: Measurement Ratios

1. $\frac{3}{12} = \frac{1}{4}$
2. $\frac{12}{12} = \frac{1}{1}$
3. $\frac{36}{12} = \frac{3}{1}$
4. $\frac{6}{3} = \frac{2}{1}$
5. $\frac{3}{3} = \frac{1}{1}$
6. $\frac{1}{3} = \frac{1}{3}$

Page 22: Comparing Inches and Feet

1. $\frac{4}{24} = \frac{1}{6}$
2. $\frac{3}{12} = \frac{1}{4}$
3. $\frac{10}{60} = \frac{1}{6}$
4. $\frac{12}{36} = \frac{1}{3}$
5. $\frac{9}{12} = \frac{3}{4}$

Page 23: Comparing Feet and Yards

1. $\frac{4}{12} = \frac{1}{3}$
2. $\frac{3}{6} = \frac{1}{2}$
3. $\frac{6}{15} = \frac{2}{5}$
4. $\frac{5}{30} = \frac{1}{6}$
5. $\frac{2}{18} = \frac{1}{9}$

Page 24: Comparing Ounces and Pounds

1. $\frac{4}{16} = \frac{1}{4}$
2. $\frac{6}{48} = \frac{1}{8}$
3. $\frac{8}{32} = \frac{1}{4}$
4. $\frac{2}{32} = \frac{1}{16}$
5. $\frac{10}{80} = \frac{1}{8}$

Page 25: Money and Time Ratios

1. $\frac{6}{20} = \frac{3}{10}$
2. $\frac{2}{20} = \frac{1}{10}$
3. $\frac{2}{8} = \frac{1}{4}$
4. $\frac{15}{60} = \frac{1}{4}$
5. $\frac{60}{100} = \frac{3}{5}$
6. $\frac{50}{120} = \frac{5}{12}$

Page 26: Ratio Review

1. a) 4 to 3
   b) 4 : 3
   c) $\frac{4}{3}$
2. $13 : 26 = \frac{13}{26} = \frac{1}{2}$
3. $48 \text{ to } 6 = \frac{48}{6} = \frac{8}{1}$
4. $\frac{3}{8} = \frac{12}{32}$
5. $\frac{8}{12} = \frac{2}{3}$
6. $\frac{45}{3} = \frac{15}{1}$

Page 26: Ratio Review (continued)

7. $\frac{25}{5} = \frac{5}{1}$ $\frac{49}{7} = \frac{7}{1}$

$5 for 1 ticket < $7 for 1 ticket

8. a) $1.35

b) $2.16

c) $2.70

9. $\frac{15}{24} = \frac{5}{8}$

10. $\frac{6}{12} = \frac{1}{2}$

11. $\frac{12}{16} = \frac{3}{4}$

12. $\frac{10}{40} = \frac{1}{4}$

13. $\frac{5}{60} = \frac{1}{12}$

CHAPTER 3: RATIO PROBLEM SOLVING

Page 27: Ratio Applications

1. $60:130 = \frac{60}{130} = \frac{6}{13}$
2. $130:20 = \frac{130}{20} = \frac{13}{2}$
3. $50:130 = \frac{50}{130} = \frac{5}{13}$
4. $130:60 = \frac{130}{60} = \frac{13}{6}$
5. $110:130 = \frac{110}{130} = \frac{11}{13}$
6. $5:12 = \frac{5}{12}$
7. $4:12 = \frac{4}{12} = \frac{1}{3}$
8. $12:4 = \frac{12}{4} = \frac{3}{1}$

Page 28: Using Ratios

1. a) $6.00
   b) $4.50

2. a) $6.00
   b) $8.00

3. a) $45.00
   b) $37.50

4. a) $40.00
   b) $15.00

Page 29: Real-Life Ratios

1. a) $18.00
   b) $3.49
   c) $12.00

2. a) $72.00
   b) $29.85
   c) $27.92

3. $8.10

4. a) $13.96
   b) $36.00
   c) $72.00

5. a) $29.85
   b) $10.47
   c) $36.00

6. $3.00

Page 30: Seeing Ratios in Word Problems

1. d) 15 : 5
2. d) 1 : 3
3. d) 74 : 121
4. b) 18 : 100

Page 31: Ratio Relationships

1. 3 : 5
2. 7 : 9
3. $.90 per pound
4. 3 : 4
5. 9 : 3 = 3 : 1
6. a) 15 : 20 = 3 : 4
   b) 5 : 20 = 1 : 4
7. 20 miles
8. $55.60

CHAPTER 4: MEANING OF PROPORTION

Page 32: What Is a Proportion?

1. $\frac{3}{4} = \frac{6}{8}$
2. $\frac{15}{5} = \frac{3}{1}$
3. $\frac{1}{4} = \frac{3}{12}$
4. $\frac{3}{2} = \frac{15}{10}$
5. $\frac{4}{7} = \frac{16}{28}$
6. $\frac{5}{6} = \frac{10}{12}$
7. $\frac{3}{4} = \frac{15}{20}$
8. $\frac{5}{10} = \frac{1}{2}$
9. $\frac{6}{27} = \frac{2}{9}$
10. $\frac{40}{100} = \frac{4}{10}$
11. $\frac{9}{7} = \frac{18}{14}$
12. $\frac{75}{100} = \frac{3}{4}$

Page 33: Simplify One Ratio

1. $\frac{4}{8} = \frac{5}{10}$
2. $\frac{6}{15} = \frac{8}{20}$
3. $\frac{3}{18} = \frac{5}{30}$
4. $\frac{8}{12} = \frac{10}{15}$
5. $\frac{9}{12} = \frac{6}{8}$
6. $\frac{18}{24} = \frac{15}{20}$
7. $\frac{18}{3} = \frac{30}{5}$
8. $\frac{15}{9} = \frac{20}{12}$
9. $\frac{7}{21} = \frac{9}{27}$
10. $\frac{10}{14} = \frac{15}{21}$
11. $\frac{6}{2} = \frac{15}{5}$
12. $\frac{30}{25} = \frac{18}{15}$

Page 34: Read the Proportion

1. $\frac{1}{4} = \frac{2}{8}$
2. $\frac{7}{3} = \frac{14}{6}$
3. $\frac{2}{5} = \frac{4}{10}$
4. $\frac{5}{8} = \frac{10}{16}$
5. $\frac{8}{12} = \frac{4}{6}$
6. $\frac{50}{25} = \frac{2}{1}$

Page 35: Two Equal Ratios

1. $\frac{2}{3} \times \frac{6}{9}$

   $2\times9 = 6\times3$

   $18 = 18$

2. $\frac{7}{3} \times \frac{21}{9}$

   $7\times9 = 21\times3$

   $63 = 63$

3. $\frac{4}{5} \times \frac{16}{20}$

   $4\times20 = 16\times5$

   $80 = 80$

4. $\frac{8}{5} \times \frac{40}{25}$

   $8\times25 = 40\times5$

   $200 = 200$

5. $\frac{6}{18} \times \frac{12}{36}$

   $6\times36 = 12\times18$

   $216 = 216$

6. $\frac{5}{1} \times \frac{10}{2}$

   $5\times2 = 10\times1$

   $10 = 10$

7. $\frac{8}{3} \times \frac{16}{6}$

   $8\times6 = 16\times3$

   $48 = 48$

8. $\frac{7}{8} \times \frac{91}{104}$

   $7\times104 = 91\times8$

   $728 = 728$

9. $\frac{10}{15} \times \frac{30}{45}$

   $10\times45 = 30\times15$

   $450 = 450$

Page 36: Cross Products

1. $\frac{1}{4} \times \frac{3}{12}$

   $1\times12 \quad 3\times4$

   $12 = 12$

2. $\frac{5}{6} \times \frac{2}{3}$

   $5\times3 \quad 2\times6$

   $15 \neq 12$

3. $\frac{21}{28} \times \frac{3}{4}$

   $21\times4 \quad 3\times28$

   $84 = 84$

4. $\frac{42}{56} \times \frac{7}{9}$

   $42\times9 \quad 7\times56$

   $378 \neq 392$

5. $\frac{9}{13} \times \frac{63}{91}$

   $9\times91 \quad 63\times13$

   $819 = 819$

6. $\frac{7}{3} \times \frac{14}{6}$

   $7\times6 \quad 14\times3$

   $42 = 42$

7. $\frac{5}{12} \times \frac{80}{192}$

   $5\times192 \quad 80\times12$

   $960 = 960$

8. $\frac{4}{7} \times \frac{52}{91}$

   $4\times91 \quad 52\times7$

   $364 = 364$

9. $\frac{12}{15} \times \frac{24}{40}$

   $12\times40 \quad 24\times15$

   $480 \neq 360$

Page 37: Proportion Readiness

1. $10 = 10$
2. $9 \neq 10$
3. $72 = 72$
4. $432 = 432$
5. $96 \neq 72$
6. $108 = 108$
7. $128 = 128$
8. $132 \neq 104$
9. $72 = 72$
10. $360 = 360$
11. $126 = 126$
12. $105 \neq 108$

Page 38: Find the Unknown Term

1. 24
2. 20
3. 2
4. 3
5. 6
6. 12

Page 39: Solve and Check

1. 20
2. 10
3. 35
4. 9
5. 8
6. 15
7. 90
8. 3
9. 4

Page 40: Apply Your Skills

1. 5
2. 12
3. 15
4. 6
5. 49
6. 45
7. 15
8. 40
9. 1,000
10. 28
11. 9
12. 3
13. 32
14. 10
15. 18

Page 41: Proportions with Fractions

1. $3\frac{1}{3}$
2. $4\frac{4}{5}$
3. $11\frac{2}{3}$
4. $10\frac{1}{2}$
5. $7\frac{7}{8}$
6. $9\frac{1}{7}$
7. $3\frac{3}{5}$
8. $15\frac{3}{4}$

Page 42: Proportions with Decimals

1. 4.8
2. 2
3. 13.09
4. 2.36
5. 35
6. 6
7. 1.6
8. 12

Page 43: Missing Term

1. $\frac{3}{4} = \frac{n}{16}$ $n = 12$
2. $\frac{4}{6} = \frac{10}{n}$ $n = 15$
3. $\frac{n}{20} = \frac{5}{8}$ $n = 12\frac{1}{2}$
4. $\frac{2.5}{n} = \frac{15}{18}$ $n = 3$

Page 43: Missing Term (continued)

5. $\frac{2.78}{2} = \frac{n}{5}$ $n = 6.95$
6. $\frac{8}{5} = \frac{13}{n}$ $n = 8\frac{1}{8}$
7. $\frac{3.25}{1} = \frac{n}{6}$ $n = 19.5$
8. $\frac{n}{4} = \frac{15}{6}$ $n = 10$

Page 44: Proportion Review

1. $\frac{5}{6} = \frac{15}{18}$
2. $\frac{4}{8} = \frac{6}{12}$
3. $\frac{9}{27} = \frac{6}{18}$
4. $7 \times 9 = 21 \times 3$
   $63 = 63$
5. $315 \neq 320$
6. $n = 60 \div 5$
   $n = 12$
7. $4 \times n = 48$
   $n = 12$
8. $54 = 5 \times n$
   $n = 10\frac{4}{5}$
9. $n \times 2.49 = 12.45$
   $n = 5$
10. $\frac{7.25}{1} = \frac{n}{6}$
    $n = 43.5$
11. $\frac{n}{4} = \frac{15}{10}$
    $n \times 10 = 60$
    $n = 6$

CHAPTER 5: PROPORTION APPLICATIONS

Page 45: Proportions in Problem Solving

1. cans to cost $\frac{5}{\$2.50}\frac{\text{cans}}{\text{cost}}$
2. cans to cost $\frac{9}{n}\frac{\text{cans}}{\text{cost}}$
3. $\frac{5}{\$2.50}\frac{\text{cans}}{\text{cost}} = \frac{9}{n}\frac{\text{cans}}{\text{cost}}$

Page 46: Setting Up Proportions

1. $\frac{3 \text{ miles}}{24 \text{ minutes}} = \frac{10 \text{ miles}}{n \text{ minutes}}$

2. $\frac{675 \text{ miles}}{n \text{ hours}} = \frac{45 \text{ miles}}{1 \text{ hour}}$

3. $\frac{2 \text{ hits}}{7 \text{ at bat}} = \frac{n \text{ hits}}{56 \text{ at bat}}$

4. $\frac{6 \text{ pounds}}{\$1.50 \text{ cost}} = \frac{13 \text{ pounds}}{n \text{ cost}}$

Page 47: Check Your Proportions

1. $\frac{3 \text{ yards}}{\$5.94 \text{ cost}} = \frac{2 \text{ yards}}{\$3.96 \text{ cost}}$

   Check: $3 \times \$3.96 = 2 \times \$5.94$
   $\$11.88 = \$11.88$

2. $\frac{255 \text{ bushels}}{3 \text{ acres}} = \frac{1{,}700 \text{ bushels}}{20 \text{ acres}}$

   Check: $255 \times 20 = 1{,}700 \times 3$
   $5{,}100 = 5{,}100$

3. $\frac{3 \text{ pounds}}{900 \text{ square feet}} = \frac{7 \text{ pounds}}{2{,}100 \text{ square feet}}$

   Check: $3 \times 2{,}100 = 7 \times 900$
   $6{,}300 = 6{,}300$

4. $\frac{120 \text{ miles}}{3 \text{ days}} = \frac{280 \text{ miles}}{7 \text{ days}}$

   Check: $120 \times 7 = 280 \times 3$
   $840 = 840$

Page 48: Using Proportions

1. $3 \times 56 = n \times 2$
   $168 = n \times 2$
   84 boys $= n$

2. $25 \times 6 = n \times 1$
   $150 = n \times 1$
   150 minutes $= n$

Page 48: Using Proportions (continued)

3. $13 \times n = 1 \times 273$
   $13 \times n = 273$
   $n = 21$ miles per gallon

4. $245 \times 7 = n \times 5$
   $1{,}715 = n \times 5$
   \$343.00 saved in 7 weeks $= n$

Page 49: Unit Prices

1. $\frac{32 \text{ ounces}}{\$2.56 \text{ cost}} = \frac{1 \text{ ounce}}{n \text{ cost}}$

   The unit price is \$.08 per ounce.

2. $\frac{6 \text{ ounces}}{\$1.32 \text{ cost}} = \frac{1 \text{ ounce}}{n \text{ cost}}$

   The unit price is \$.22 per ounce.

3. $\frac{\$2.64 \text{ cost}}{24 \text{ tablets}} = \frac{n \text{ cost}}{1 \text{ tablet}}$

   The unit price is \$.11 per tablet.

4. $\frac{\$1.96 \text{ cost}}{7 \text{ ounces}} = \frac{n \text{ cost}}{1 \text{ ounce}}$

   The unit price is \$.28 per ounce.

Page 50: Find the Unit Costs

1. $\frac{6 \text{ pounds}}{2.88 \text{ dollars}} = \frac{1 \text{ pound}}{n \text{ dollars}}$

   Unit price: \$.48 per pound

2. $\frac{8 \text{ pounds}}{2.16 \text{ dollars}} = \frac{1 \text{ pound}}{n \text{ dollars}}$

   Unit price: \$.27 per pound

3. $\frac{5 \text{ grapefruit}}{2.85 \text{ dollars}} = \frac{1 \text{ grapefruit}}{n \text{ dollars}}$

   Unit price: \$.57 each

Page 50: Find the Unit Costs (continued)

4. $\frac{3 \text{ pounds}}{1.44 \text{ dollars}} = \frac{1 \text{ pound}}{n \text{ dollars}}$

Unit price: $.48 per pound

5. $\frac{16 \text{ ounces}}{.32 \text{ dollars}} = \frac{1 \text{ ounce}}{n \text{ dollars}}$

Unit price: $.02 per ounce

6. $\frac{7 \text{ apples}}{1.40 \text{ dollars}} = \frac{1 \text{ apple}}{n \text{ dollars}}$

Unit price: $.20 each

Page 51: Comparison Shopping

1. $\frac{9 \text{ ounces}}{2.70 \text{ dollars}} = \frac{1 \text{ ounce}}{n \text{ dollars}}$

The unit price is $.30 per ounce.

$\frac{15 \text{ ounces}}{3.75 \text{ dollars}} = \frac{1 \text{ ounce}}{n \text{ dollars}}$

The unit price is $.25 per ounce.

The 15-ounce size is cheaper per ounce.

2. $\frac{6 \text{ bars}}{2.28 \text{ dollars}} = \frac{1 \text{ bar}}{n \text{ dollars}}$

The unit price is $.38 per candy bar.

$\frac{3 \text{ bars}}{1.08 \text{ dollars}} = \frac{1 \text{ bar}}{n \text{ dollars}}$

The unit price is $.36 per candy bar.

The package containing 3 candy bars is cheaper per bar.

3. $\frac{200 \text{ feet}}{6.00 \text{ dollars}} = \frac{1 \text{ foot}}{n \text{ dollars}}$

The unit price is $.03 per foot.

$\frac{25 \text{ feet}}{1.00 \text{ dollars}} = \frac{1 \text{ foot}}{n \text{ dollars}}$

The unit price is $.04 per foot.

The 200-foot size is cheaper per foot.

Page 51: Comparison Shopping (continued)

4. $\frac{16 \text{ ounces}}{2.56 \text{ dollars}} = \frac{1 \text{ ounce}}{n \text{ dollars}}$

The unit price is $.16 per ounce.

$\frac{6 \text{ ounces}}{1.32 \text{ dollars}} = \frac{1 \text{ ounce}}{n \text{ dollars}}$

The unit price is $.22 per ounce.

The 16-ounce size is cheaper per ounce.

CHAPTER 6: PROPORTION PROBLEM SOLVING

Page 52: Changing Recipes

A. $5 = n$
5 eggs are needed to make 25 waffles.

1. $\frac{2 \text{ eggs}}{24 \text{ cookies}} = \frac{n \text{ eggs}}{60 \text{ cookies}}$

5 eggs are needed to make 60 cookies.

2. $\frac{3 \text{ ounces}}{9 \text{ servings}} = \frac{4 \text{ ounces}}{n \text{ servings}}$

4 ounces of cream cheese will make 12 servings.

3. $\frac{12 \text{ people}}{4 \text{ oranges}} = \frac{18 \text{ people}}{n \text{ oranges}}$

6 oranges are needed to make punch for 18 people.

4. $\frac{3 \text{ teaspoons}}{36 \text{ servings}} = \frac{n \text{ teaspoons}}{12 \text{ servings}}$

1 teaspoon of butter is needed for 12 servings.

5. $\frac{6 \text{ tablespoons}}{4 \text{ servings}} = \frac{9 \text{ tablespoons}}{n \text{ servings}}$

9 tablespoons of milk will make 6 servings.

Page 52: Changing Recipes (continued)

6. $\frac{6 \text{ teaspoons}}{30 \text{ biscuits}} = \frac{n \text{ teaspoons}}{40 \text{ biscuits}}$

   8 teaspoons of baking powder are needed to make 40 biscuits.

Page 53: Figuring Costs

A. $n = \$9.90$
   Five boxes of cereal cost $9.90.

1. $\frac{4.80 \text{ dollars}}{1 \text{ square yard}} = \frac{n \text{ dollars}}{8 \text{ square yards}}$

   8 square yards will cost $38.40.

2. $\frac{2 \text{ notebooks}}{2.34 \text{ dollars}} = \frac{n \text{ notebooks}}{7.02 \text{ dollars}}$

   Patricia could buy 6 notebooks for $7.02.

3. $\frac{3 \text{ candy bars}}{2.25 \text{ dollars}} = \frac{10 \text{ candy bars}}{n \text{ dollars}}$

   10 candy bars would cost $7.50.

4. $\frac{6 \text{ pounds}}{1.50 \text{ dollars}} = \frac{13 \text{ pounds}}{n \text{ dollars}}$

   13 pounds of potatoes will cost $3.25.

5. $\frac{3 \text{ cans}}{1.26 \text{ dollars}} = \frac{9 \text{ cans}}{n \text{ dollars}}$

   Sherlene spent $3.78 for 9 cans of soup.

6. $\frac{1.59 \text{ dollars}}{3 \text{ oranges}} = \frac{n \text{ dollars}}{7 \text{ oranges}}$

   7 oranges will cost $3.71.

Page 54: Travel Plans

A. $n = 7$
   It will take Kevin 7 hours.

1. $\frac{21 \text{ miles}}{1 \text{ gallon}} = \frac{315 \text{ miles}}{n \text{ gallons}}$

   15 gallons of gasoline will be used on a 315-mile trip.

Page 54: Travel Plans (continued)

2. $\frac{1{,}365 \text{ miles}}{3 \text{ hours}} = \frac{n \text{ miles}}{5 \text{ hours}}$

   A jet can fly 2,275 miles in 5 hours.

3. $\frac{55 \text{ miles}}{1 \text{ hour}} = \frac{220 \text{ miles}}{n \text{ hours}}$

   It would take the train 4 hours to travel 220 miles.

4. $\frac{45 \text{ miles}}{1 \text{ hour}} = \frac{n \text{ miles}}{4.5 \text{ hours}}$

   The train will travel 202.5 miles in 4.5 hours.

5. $\frac{.35 \text{ dollars}}{25 \text{ miles}} = \frac{n \text{ dollars}}{155 \text{ miles}}$

   It would cost $2.17 to travel 155 miles.

6. $\frac{104 \text{ miles}}{2 \text{ hours}} = \frac{260 \text{ miles}}{n \text{ hours}}$

   It will take Benjamin 5 hours to travel 260 miles.

Page 55: Scale Drawings

1. $\frac{1 \text{ inch (map)}}{170 \text{ miles (actual)}} = \frac{2.5 \text{ inches (map)}}{n \text{ miles (actual)}}$

   $n = 425$

   The actual distance between El Paso and San Antonio is 425 miles.

2. $\frac{1 \text{ inch (map)}}{170 \text{ miles (actual)}} = \frac{1.5 \text{ inches (map)}}{n \text{ miles (actual)}}$

   $n = 255$

   The actual distance between Lubbock and Dallas is 255 miles.

3. $\frac{1 \text{ inch (map)}}{170 \text{ miles (actual)}} = \frac{2 \text{ inches (map)}}{n \text{ miles (actual)}}$

   $n = 340$

   The actual distance between Wichita Falls and Corpus Christi is 340 miles.

Page 55: Scale Drawings (continued)

4. $\frac{1 \text{ inch (map)}}{170 \text{ miles (actual)}} = \frac{.5 \text{ inch (map)}}{n \text{ miles (actual)}}$

$n = 85$

The actual distance between Texarkana and Tyler is 85 miles.

Page 56: Map Applications

1. $\frac{1 \text{ inch}}{150 \text{ miles}} = \frac{4 \text{ inches}}{n \text{ miles}}$

The two cities are 600 miles apart.

2. $\frac{1 \text{ inch}}{45 \text{ miles}} = \frac{n \text{ inches}}{315 \text{ miles}}$

7 inches on the map represent 315 miles.

3. $\frac{1 \text{ inch}}{75 \text{ miles}} = \frac{3 \text{ inches}}{n \text{ miles}}$

3 inches on the map represent 225 miles.

4. $\frac{2 \text{ inches}}{150 \text{ miles}} = \frac{5 \text{ inches}}{n \text{ miles}}$

5 inches on the map represent 375 miles.

5. $\frac{2 \text{ inches}}{150 \text{ miles}} = \frac{n \text{ inches}}{375 \text{ miles}}$

375 miles are represented by 5 inches.

6. $\frac{1.5 \text{ inches}}{20 \text{ miles}} = \frac{6 \text{ inches}}{n \text{ miles}}$

6 inches on the map represent 80 miles.

7. $\frac{1 \text{ inch}}{150 \text{ miles}} = \frac{3.5 \text{ inches}}{n \text{ miles}}$

It is 525 miles between two cities that are 3.5 inches apart.

8. $\frac{.5 \text{ inch}}{50 \text{ miles}} = \frac{7 \text{ inches}}{n \text{ miles}}$

7 inches on the scale equal 700 miles.

Page 57: Make a Chart

1.

hours of sleep	9	n
days	1	365

2.

dollars earned	10	155
dollars saved	1	n

3.

inches of snow	1.5	n
hours	1	4

4.

swimmers	6	n
number of kids	10	350

Page 58: Turn Charts into Proportions

1.

dollar bills	1	n
dimes	10	130

$\frac{1}{10} = \frac{n}{130}$

$n = 13$ dollar bills

2.

tapes bought	6	9
free tapes	2	n

$\frac{6}{2} = \frac{9}{n}$

$n = 3$ free tapes

3.

boxes of cereal	1.5	n
bags of pretzels	1	3

$\frac{1.5}{1} = \frac{n}{3}$

$n = 4.5$ boxes of cereal needed

4.

aspirin tablets	2	n
hours	4	24

$\frac{2}{4} = \frac{n}{24}$

$n = 12$ aspirin tablets in 24 hours

Page 59: Proportion in Measurement

1.

pound	1	n
ounces	16	128

$$\frac{1 \text{ pound}}{16 \text{ ounces}} = \frac{n \text{ pounds}}{128 \text{ ounces}}$$

n = 8 pounds

2.

pecks	4	n
bushels	1	52

$$\frac{4 \text{ pecks}}{1 \text{ bushel}} = \frac{n \text{ pecks}}{52 \text{ bushels}}$$

n = 208 pecks

3.

grams	1	564
ounces	.04	n

$$\frac{1 \text{ gram}}{.04 \text{ ounce}} = \frac{564 \text{ grams}}{n \text{ ounces}}$$

n = 22.56 ounces

4.

centimeters	100	n
meters	1	655

$$\frac{100 \text{ centimeters}}{1 \text{ meter}} = \frac{n \text{ centimeters}}{655 \text{ meters}}$$

n = 65,500 centimeters

Page 60: Proportions in Sports

1.

miles	120	n
days	3	7

$$\frac{120 \text{ miles}}{3 \text{ days}} = \frac{n \text{ miles}}{7 \text{ days}}$$

n = 280 miles

Page 60: Proportions in Sports (continued)

2.

miles	8.4	n
hours	2	3

$$\frac{8.4 \text{ miles}}{2 \text{ hours}} = \frac{n \text{ miles}}{3 \text{ hours}}$$

n = 12.6 miles

3.

home runs	4	n
games	6	9

$$\frac{4 \text{ home runs}}{6 \text{ games}} = \frac{n \text{ home runs}}{9 \text{ games}}$$

n = 6 home runs

4.

points	92	n
games	4	18

$$\frac{92 \text{ points}}{4 \text{ games}} = \frac{n \text{ points}}{18 \text{ games}}$$

n = 414 points

CHAPTER 1: MEANING OF PERCENT

Page 1: What Is Percent?

1. 45
2. $\frac{45}{100}$
3. 45%
4. 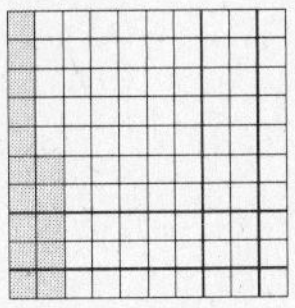
5. $\frac{15}{100}$
6. 15%
7. 25
8. 15
9. 75
10. 90
11. 5
12. 1

Page 2: Shade the Percents

1. 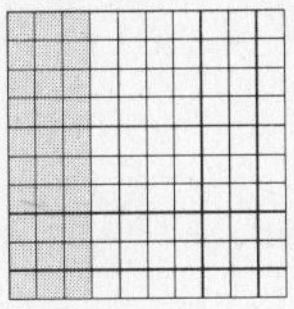
2. $\frac{30}{100}$
3. 30%
4. 70%
5. 100%
6. 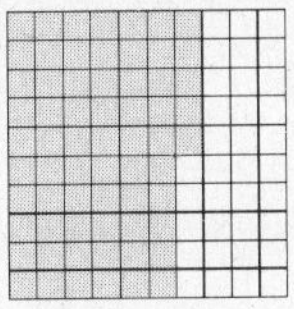
7. $\frac{65}{100}$
8. 65%
9. 35%
10. 100%
11. 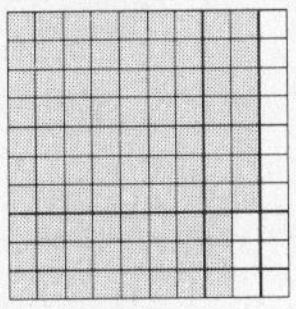
12. $\frac{87}{100}$
13. 87%
14. 13%
15. 100%

Page 3: Understanding Percent

1. 45 out of 100 or $\frac{45}{100}$
2. 60 out of 100 or $\frac{60}{100}$
3. 55 out of 100 or $\frac{55}{100}$
4. 5 out of 100 or $\frac{5}{100}$
5. $\frac{95}{100}$ = 95%
6. $\frac{100}{100}$ = 100%
7. $\frac{50}{100}$ = 50%
8. $\frac{75}{100}$ = 75%
9. 33 out of 100 or 33%
10. 20 out of 100 or 20%
11. 10 out of 100 or 10%
12. 18 out of 100 or 18%

Page 4: Percent Means Parts Out of 100

1. $\frac{24}{100}$ = 24%
2. $\frac{52}{100}$ = 52%
3. $\frac{48}{100}$ = 48%
4. $\frac{69}{100}$ = 69%
5. $\frac{33}{100}$ = 33%
6. $\frac{60}{100}$ = 60%

Page 5: Shade the Squares

1. 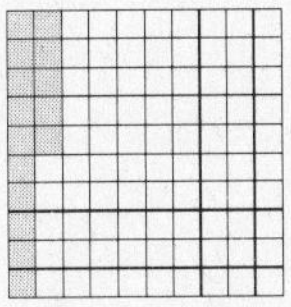

$\frac{15}{100}$ = 15%

2. 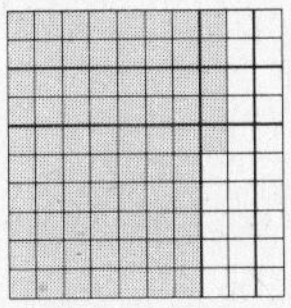

$\frac{75}{100}$ = 75%

3. 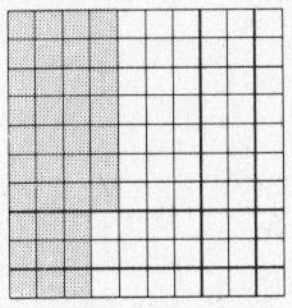

$\frac{37}{100}$ = 37%

4. 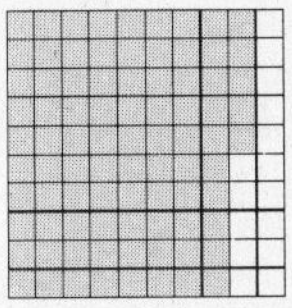

$\frac{85}{100}$ = 85%

5. 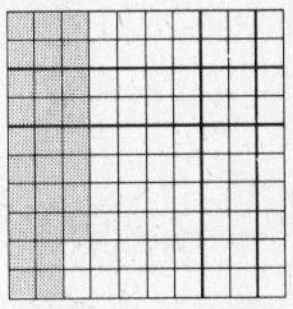

$\frac{28}{100}$ = 28%

6. 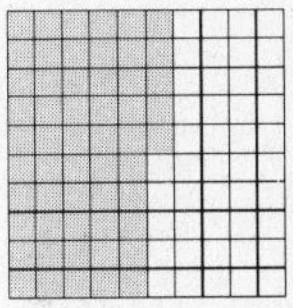

$\frac{55}{100}$ = 55%

Page 6: Number Lines

1. 70% 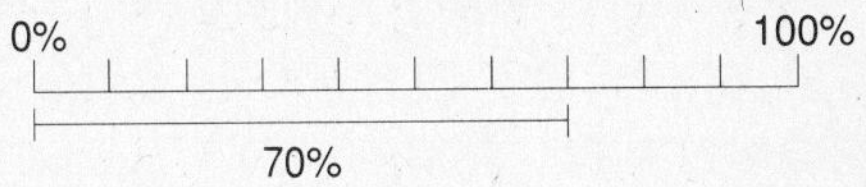

2. 15% 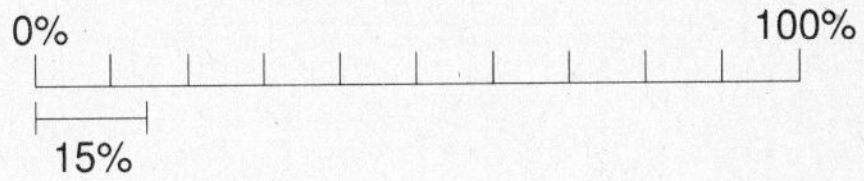

3. 95% 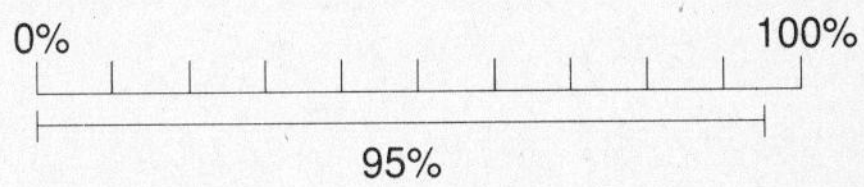

4. 35% 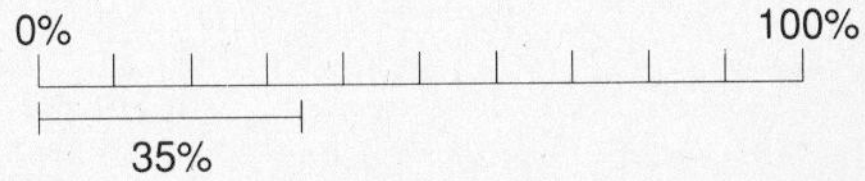

5. 20% 

Page 7: One Hundred Percent Equals One

A. 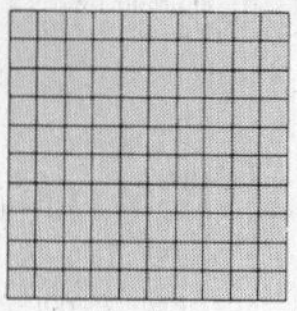

B. 100%
C. 100%

1. 50%
2. 25%
3. 91%
4. 55%
5. 36%
6. 45%
7. 30%
8. 40%
9. 55%

Page 8: Fifty Percent

A. $\frac{50}{100} = 50\%$

1. a) 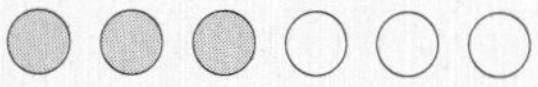
   b) 3
   c) 3
2. a) 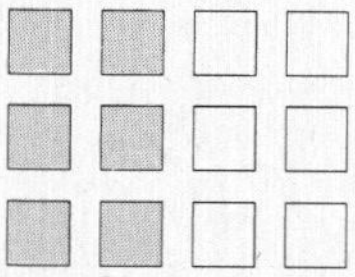
   b) 6
   c) 6
3. a) 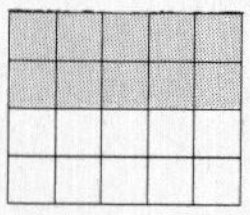
   b) 10
   c) 10
4. 5
5. 8
6. 15
7. 12
8. 9
9. 25
10. 20

Page 9: Twenty-Five Percent

A. $\frac{25}{100} = 25\%$

1. a) 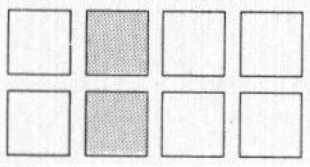
   b) 2
   c) 2
2. a) 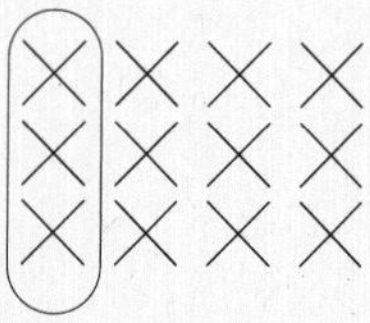
   b) 3
   c) 3
3. 8
4. 4
5. 5
6. 10
7. 6
8. 9
9. 1
10. 7

Page 10: What Does the Percent Mean?

1. 1 out of 2
2. 5 out of 10
3. 10 out of 20
4. 15 out of 30
5. 9 out of 18
6. 25 out of 50
7. 1 out of 4
8. 2 out of 8
9. 5 out of 20
10. 10 out of 40
11. 3 out of 12
12. 4 out of 16
13. 1 out of 10
14. 3 out of 30
15. 2 out of 20
16. 8 out of 80
17. 5 out of 50
18. 9 out of 90
19. 1 out of 5
20. 5 out of 25
21. 2 out of 10
22. 7 out of 35
23. 3 out of 15
24. 9 out of 45

Page 11: More Percent Meanings

1. 3 out of 5
2. 6 out of 10
3. 12 out of 20
4. 24 out of 40
5. 30 out of 50
6. 36 out of 60
7. 2 out of 5
8. 20 out of 50
9. 4 out of 10
10. 16 out of 40
11. 8 out of 20
12. 12 out of 30
13. 4 out of 5
14. 8 out of 10
15. 12 out of 15
16. 20 out of 25
17. 36 out of 45
18. 28 out of 35
19. 3 out of 4
20. 6 out of 8
21. 15 out of 20
22. 21 out of 28
23. 9 out of 12
24. 18 out of 24

CHAPTER 2: FRACTIONS AND PERCENTS

Page 12: Common Fractions and Percents

1. $33\frac{1}{3}\%$
2. 75%
3. 25%
4. 50%
5. 10%
6. 50%
7. 25%
8. $66\frac{2}{3}\%$
9. $33\frac{1}{3}\%$

Page 13: Rename as a Percent

1. $\frac{1}{2} = \frac{50}{100} = 50\%$
2. $\frac{1}{4} = \frac{25}{100} = 25\%$
3. $\frac{1}{5} = \frac{20}{100} = 20\%$
4. $\frac{1}{10} = \frac{10}{100} = 10\%$

Page 14: Ratios to Percents

1. 3
2. $\frac{3}{10}$
3. 30; $\frac{30}{100} = 30\%$
4. 7
5. 70%
6. 5
7. $\frac{5}{20}$
8. 25; $\frac{25}{100} = 25\%$
9. 15
10. 75%

Page 15: Apply Your Skills

1. $\frac{1 \times 50}{2 \times 50} = \frac{50}{100} = 50\%$
2. $\frac{1 \times 25}{4 \times 25} = \frac{25}{100} = 25\%$
3. $\frac{4 \times 10}{10 \times 10} = \frac{40}{100} = 40\%$
4. $\frac{2 \times 20}{5 \times 20} = \frac{40}{100} = 40\%$

Page 16: Use the Pictures

1. $\frac{1}{4} = \frac{25}{100} = 25\%$
2. $\frac{3}{4} = \frac{75}{100} = 75\%$
3. $\frac{4}{5} = \frac{80}{100} = 80\%$
4. $\frac{3}{5} = \frac{60}{100} = 60\%$
5. $\frac{6}{8} = \frac{3}{4} = \frac{75}{100} = 75\%$
6. $\frac{2}{8} = \frac{1}{4} = \frac{25}{100} = 25\%$
7. $\frac{3}{6} = \frac{1}{2} = \frac{50}{100} = 50\%$
8. $\frac{4}{16} = \frac{1}{4} = \frac{25}{100} = 25\%$

Page 17: Shade the Percents

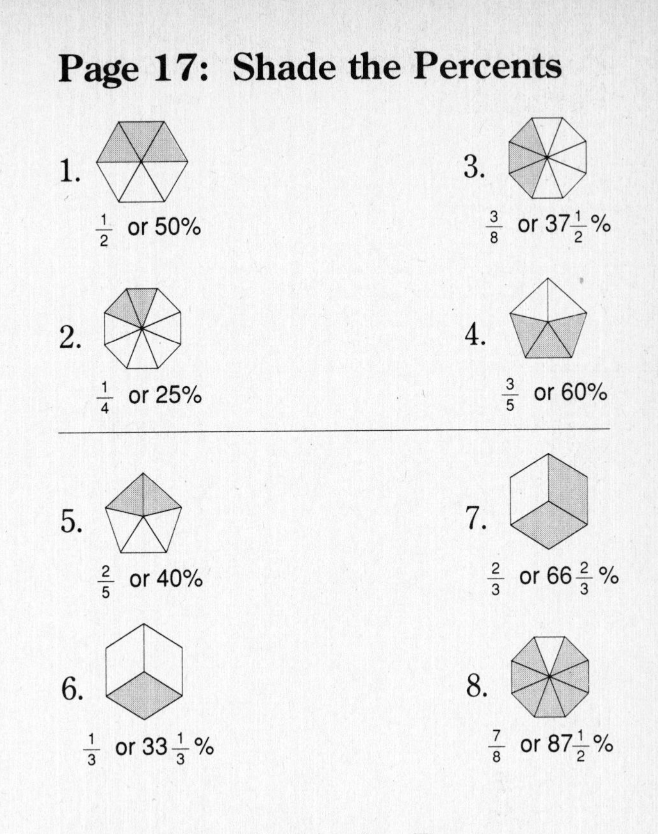

Page 18: Name the Percent

1. $\frac{7}{10} = \frac{70}{100} = 70\%$
2. $\frac{4}{20} = \frac{20}{100} = 20\%$
3. $\frac{4}{5} = \frac{80}{100} = 80\%$
4. $\frac{10}{25} = \frac{40}{100} = 40\%$

Page 19: Think It Through

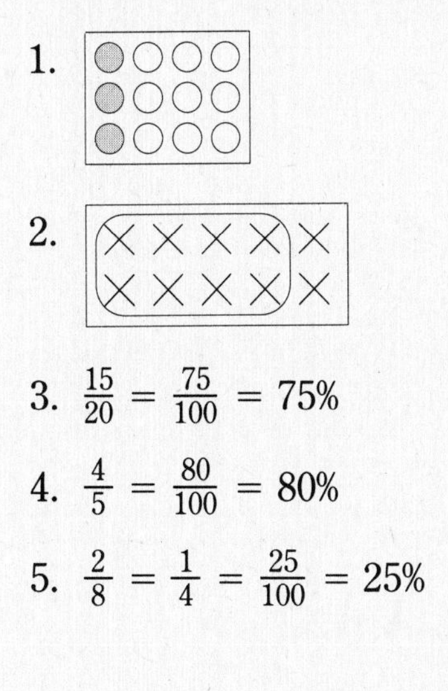

3. $\frac{15}{20} = \frac{75}{100} = 75\%$
4. $\frac{4}{5} = \frac{80}{100} = 80\%$
5. $\frac{2}{8} = \frac{1}{4} = \frac{25}{100} = 25\%$

Page 20: Change Fractions to Percents

A. 50%

B. 40%

C. 75% $\frac{3}{4} \times \frac{25}{25} = \frac{75}{100} = 75\%$

1. $\frac{2}{10} = \frac{20}{100} = 20\%$
2. $\frac{3}{5} = \frac{60}{100} = 60\%$
3. $\frac{1}{4} = \frac{25}{100} = 25\%$
4. $\frac{7}{10} = \frac{70}{100} = 70\%$
5. $\frac{1}{25} = \frac{4}{100} = 4\%$
6. $\frac{4}{5} = \frac{80}{100} = 80\%$
7. $\frac{1}{20} = \frac{5}{100} = 5\%$
8. $\frac{5}{5} = \frac{100}{100} = 100\%$
9. $\frac{7}{20} = \frac{35}{100} = 35\%$
10. $\frac{9}{50} = \frac{18}{100} = 18\%$

Page 21: Percent Wise

1. 80%
2. 20%
3. 30%
4. 70%
5. 65%
6. 35%
7. 75%
8. 25%

Page 22: Show What Percent Is Shaded

1. a) 50%
   b) 50%
2. a) 75%
   b) 25%
3. a) 60%
   b) 40%
4. a) 60%
   b) 40%
5. a) 100%
   b) 0%
6. a) 36%
   b) 64%

Page 23: Rename Percents as Fractions

1. $75\% = \frac{75}{100} = \frac{3}{4}$
2. $80\% = \frac{80}{100} = \frac{4}{5}$
3. $60\% = \frac{60}{100} = \frac{3}{5}$
4. $10\% = \frac{10}{100} = \frac{1}{10}$

Page 24: Percents Are Special Ratios

1. a) 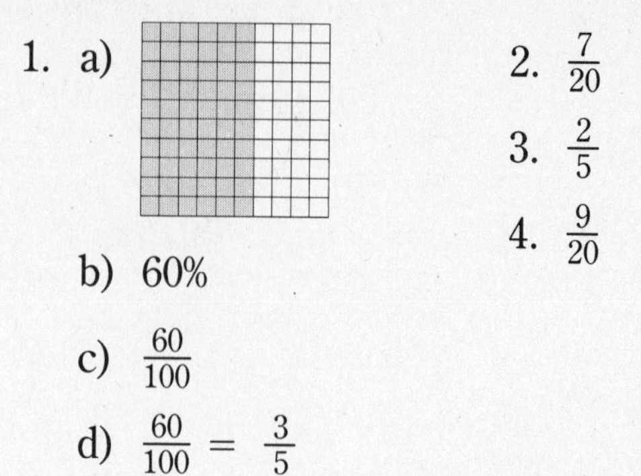

   b) 60%

   c) $\frac{60}{100}$

   d) $\frac{60}{100} = \frac{3}{5}$
2. $\frac{7}{20}$
3. $\frac{2}{5}$
4. $\frac{9}{20}$

Page 25: Change Percents to Fractions

1. $\frac{1}{5}$
2. $\frac{1}{10}$
3. $\frac{30}{100} = \frac{3}{10}$
4. $\frac{50}{100} = \frac{1}{2}$
5. $\frac{70}{100} = \frac{7}{10}$
6. $\frac{4}{100} = \frac{1}{25}$
7. $\frac{90}{100} = \frac{9}{10}$
8. $\frac{3}{4}$
9. $\frac{1}{20}$
10. $\frac{45}{100} = \frac{9}{20}$
11. $\frac{80}{100} = \frac{4}{5}$
12. $\frac{15}{100} = \frac{3}{20}$
13. $\frac{60}{100} = \frac{3}{5}$
14. $\frac{40}{100} = \frac{2}{5}$

Page 26: Practice Changing Percents to Fractions

1. $\frac{25}{100} = \frac{1}{4}$
2. $\frac{8}{100} = \frac{2}{25}$
3. $\frac{15}{100} = \frac{3}{20}$
4. $\frac{4}{100} = \frac{1}{25}$
5. $\frac{75}{100} = \frac{3}{4}$
6. $\frac{20}{100} = \frac{1}{5}$
7. $\frac{90}{100} = \frac{9}{10}$
8. $\frac{6}{100} = \frac{3}{50}$
9. $\frac{12}{100} = \frac{3}{25}$
10. $\frac{10}{100} = \frac{1}{10}$
11. $\frac{45}{100} = \frac{9}{20}$
12. $\frac{1}{100} = \frac{1}{100}$
13. $\frac{17}{100} = \frac{17}{100}$
14. $\frac{22}{100} = \frac{11}{50}$
15. $\frac{60}{100} = \frac{3}{5}$
16. $\frac{36}{100} = \frac{9}{25}$
17. $\frac{9}{100} = \frac{9}{100}$
18. $\frac{28}{100} = \frac{7}{25}$

Page 27: Common Equivalents

1. $\frac{1}{4}$
2. $\frac{1}{5}$
3. $\frac{1}{2}$
4. $\frac{1}{10}$
5. $\frac{2}{3}$
6. $\frac{1}{8}$
7. $\frac{3}{4}$
8. $\frac{1}{3}$
9. $66\frac{2}{3}\%$
10. $12\frac{1}{2}\%$
11. 75%
12. 20%
13. 10%
14. 25%
15. $33\frac{1}{3}\%$
16. 50%

CHAPTER 3: PERCENTS GREATER THAN 100

Page 28: Numbers Greater Than 100%

1.

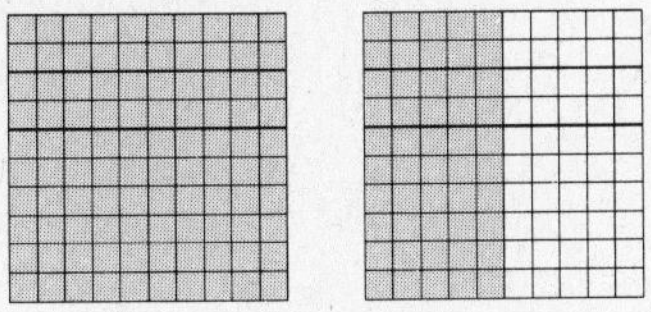

100% + 50% = 150%

$1 + \frac{1}{2} = 1\frac{1}{2}$

2.

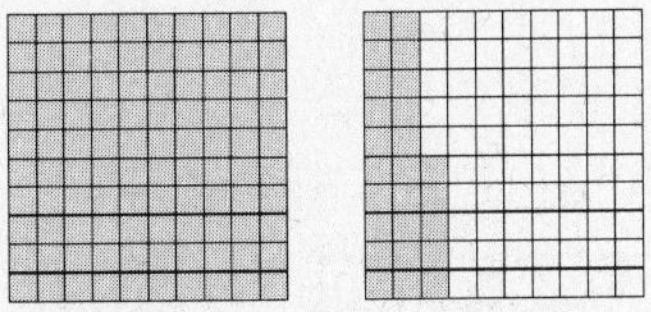

100% + 25% = 125%

$1 + \frac{1}{4} = 1\frac{1}{4}$

3.

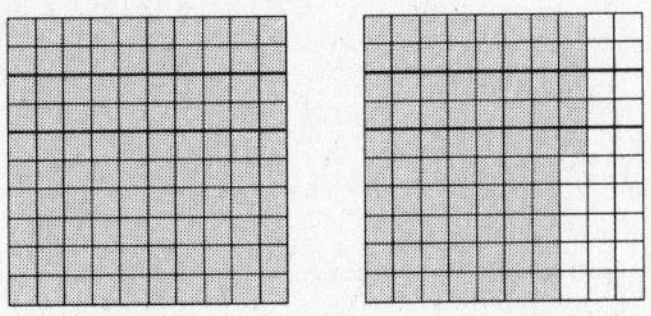

100% + 75% = 175%

$1 + \frac{3}{4} = 1\frac{3}{4}$

Page 28: Numbers Greater Than 100% (continued)

4.

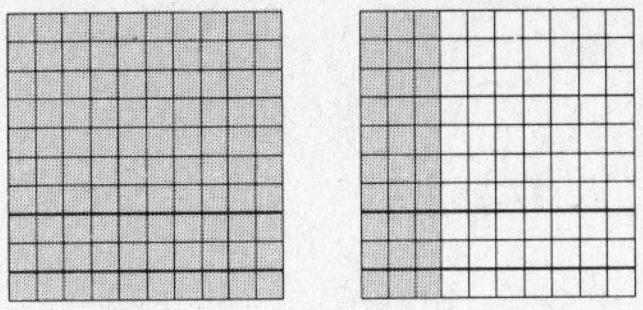

100% + 30% = 130%

$1 + \frac{3}{10} = 1\frac{3}{10}$

5.

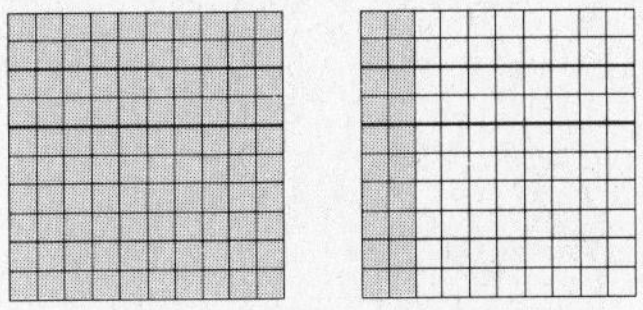

100% + 20% = 120%

$1 + \frac{1}{5} = 1\frac{1}{5}$

6.

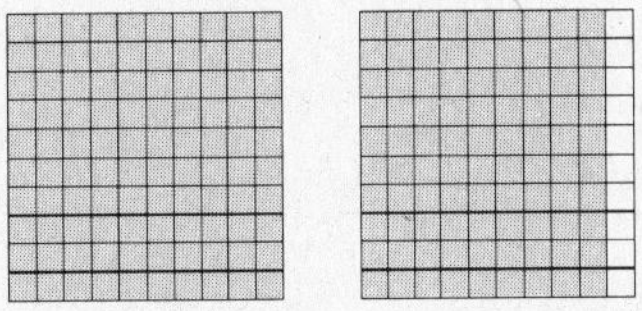

100% + 90% = 190%

$1 + \frac{9}{10} = 1\frac{9}{10}$

Page 29: Change Percents to Mixed Numbers

1. $1 = \frac{100}{100} = 100\%$
2. $2 = \frac{200}{100} = 200\%$
3. $5 = \frac{500}{100} = 500\%$
4. $3 = \frac{300}{100} = 300\%$
5. $4 = \frac{400}{100} = 400\%$
6. $8 = \frac{800}{100} = 800\%$
7. $310\% = 300\% + 10\%$

$= 3 + \frac{10}{100}$

$= 3 + \frac{1}{10} = 3\frac{1}{10}$

Page 29: Change Percents to Mixed Numbers (continued)

8. $450\% = 400\% + 50\%$

$= 4 + \frac{50}{100}$

$= 4 + \frac{1}{2} = 4\frac{1}{2}$

9. $225\% = 200\% + 25\%$

$= 2 + \frac{25}{100}$

$= 2\frac{1}{4}$

10. $460\% = 400\% + 60\%$

$= 4 + \frac{60}{100}$

$= 4\frac{3}{5}$

11. $270\% = 200\% + 70\%$

$= 2 + \frac{70}{100}$

$= 2\frac{7}{10}$

12. $580\% = 500\% + 80\%$

$= 5 + \frac{80}{100}$

$= 5\frac{4}{5}$

Page 30: Mixed Practice

1. $\frac{20}{100} = \frac{1}{5}$
2. $2 + \frac{75}{100} = 2\frac{3}{4}$
3. $3 + \frac{50}{100} = 3\frac{1}{2}$
4. $\frac{45}{100} = \frac{9}{20}$
5. $\frac{5}{100} = \frac{1}{20}$
6. $1 + \frac{60}{100} = 1\frac{3}{5}$
7. $\frac{8}{100} = \frac{2}{25}$
8. $6\frac{1}{4}$
9. $\frac{1}{10}$
10. $4\frac{1}{5}$
11. $\frac{4}{5}$
12. $\frac{1}{25}$
13. $1\frac{1}{20}$
14. $\frac{17}{20}$

Page 31: More Mixed Practice

1. 15%
2. 25%
3. 70%
4. 25%
5. $2\frac{3}{4}$
6. 10%
7. $\frac{3}{10}$
8. 70%
9. 750%
10. 50%
11. 20%
12. 40%
13. 50%
14. $\frac{9}{10}$
15. 25%

Page 32: Percent Review

1. X X X X
   X X X X
2. 50%
3. 25%
4. 75%
5. 20%
6. 610%
7. 9
8. 8
9. $\frac{1}{5}$
10. $\frac{1}{2}$
11. $2\frac{1}{2}$
12. $\frac{1}{10}$
13. 80%
14. 20%
15. 50%
16. 100%
17. 10%
18. 25%

CHAPTER 4: PERCENTS AND DECIMALS

Page 33: Change Percents to Decimals

1. 10 parts out of 100 = $\frac{10}{100}$ = .10
2. 85 parts out of 100 = $\frac{85}{100}$ = .85
3. 33 parts out of 100 = $\frac{33}{100}$ = .33
4. 40 parts out of 100 = $\frac{40}{100}$ = .40
5. 15 parts out of 100 = $\frac{15}{100}$ = .15
6. 5 parts out of 100 = $\frac{5}{100}$ = .05

Page 33: Change Percents to Decimals (continued)

7. 3 parts out of 100 = $\frac{3}{100}$ = .03
8. 1 part out of 100 = $\frac{1}{100}$ = .01

Page 34: Change to Hundredths

1. $\frac{15}{100}$ = .15
2. $\frac{18}{100}$ = .18
3. $\frac{39}{100}$ = .39
4. $\frac{7}{100}$ = .07
5. $\frac{20}{100}$ = .20
6. $\frac{9}{100}$ = .09
7. $\frac{75}{100}$ = .75
8. $\frac{65}{100}$ = .65
9. $\frac{5}{100}$ = .05
10. $\frac{98}{100}$ = .98
11. $\frac{67}{100}$ = .67
12. $\frac{12}{100}$ = .12
13. $\frac{10}{100}$ = .10
14. $\frac{4}{100}$ = .04

Page 35: Write Percents as Decimals

1. .75
2. .36
3. .16
4. .44
5. .70
6. .93
7. .08
8. .82
9. .10
10. .05
11. .07
12. .22

Page 36: Change Percents to Mixed Decimals

1. 3.25
2. 4.01
3. 1.36
4. 2.19
5. 1.50
6. 6.00
7. 3.00
8. 2.75
9. 1.00
10. 1.92
11. 5.00
12. 4.50

Page 37: Mixed Practice

1. .054
2. .217
3. .0368
4. .45
5. 3.00
6. .095
7. 3.48
8. .026
9. 3.50
10. .016
11. .543
12. .19

Page 38: Change Decimals to Percents

1. $\frac{65}{100}$ = 65 hundredths = 65%
2. $\frac{19}{100}$ = 19 hundredths = 19%
3. $\frac{33}{100}$ = 33 hundredths = 33%
4. $\frac{80}{100}$ = 80 hundredths = 80%
5. $\frac{1}{100}$ = 1 hundredth = 1%
6. $\frac{68}{100}$ = 68 hundredths = 68%
7. $\frac{75}{100}$ = 75 hundredths = 75%
8. $\frac{66}{100}$ = 66 hundredths = 66%
9. $\frac{8}{100}$ = 8 hundredths = 8%
10. $\frac{86}{100}$ = 86 hundredths = 86%

Page 39: Change One-Place Decimals to Percents

1. $\frac{8}{10} \times \frac{10}{10} = \frac{80}{100}$ = 80%
2. $\frac{1}{10} \times \frac{10}{10} = \frac{10}{100}$ = 10%
3. $\frac{6}{10} \times \frac{10}{10} = \frac{60}{100}$ = 60%
4. $\frac{5}{10} \times \frac{10}{10} = \frac{50}{100}$ = 50%
5. $\frac{3}{10} \times \frac{10}{10} = \frac{30}{100}$ = 30%
6. $\frac{2}{10} \times \frac{10}{10} = \frac{20}{100}$ = 20%
7. $\frac{7}{10} \times \frac{10}{10} = \frac{70}{100}$ = 70%
8. $\frac{9}{10} \times \frac{10}{10} = \frac{90}{100}$ = 90%

Page 40: Relating Decimals to Percents

1. 26%
2. 35.4%
3. 148%
4. 88.7%
5. 36%
6. 10.9%
7. 44%
8. 15.6%
9. 78%
10. 292%
11. 57%
12. 113%

Page 41: Add Zeros When Necessary

1. 30%
2. 50%
3. 340%
4. 910%
5. 70%
6. 800%
7. 80%
8. 600%
9. 40%
10. 590%
11. 10%
12. 320%

CHAPTER 5: DECIMALS, FRACTIONS, AND PERCENTS

Page 42: Put It All Together

1. a) $\frac{1}{2}$
   b) .50
   c) 50%
2. a) $\frac{3}{4}$
   b) .75
   c) 75%
3. a) $\frac{1}{5}$
   b) .20
   c) 20%
4. a) $\frac{7}{10}$
   b) .70
   c) 70%
5. a) $\frac{4}{5}$
   b) .80
   c) 80%
6. a) $\frac{3}{5}$
   b) .60
   c) 60%

Page 43: Equivalents

	Fraction (Simplified)	Fraction (Hundredths)	Decimal	Percent
1.	$\frac{1}{2}$	$\frac{50}{100}$	.50	50%
2.	$\frac{1}{4}$	$\frac{25}{100}$	.25	25%
3.	$\frac{3}{4}$	$\frac{75}{100}$	.75	75%
4.	$\frac{1}{10}$	$\frac{10}{100}$	.10	10%
5.	$\frac{7}{10}$	$\frac{70}{100}$	.70	70%
6.	$\frac{3}{5}$	$\frac{60}{100}$	.60	60%
7.	$\frac{2}{5}$	$\frac{40}{100}$	.40	40%
8.	$\frac{7}{100}$	$\frac{7}{100}$	.07	7%
9.	$\frac{1}{50}$	$\frac{2}{100}$	.02	2%
10.	$\frac{1}{5}$	$\frac{20}{100}$	.20	20%

Page 44: Comparing Percents, Fractions, and Decimals

1. =
2. <
3. <
4. <
5. <
6. <
7. <
8. >
9. =
10. >
11. <
12. =

Page 45: Using Proportions

A. 20%
Step 3: 200
Step 4: 200 $66\frac{2}{3} = n$
B. $66\frac{2}{3}\%$

Page 46: Use Cross Products

1. $33\frac{1}{3}\%$
2. $16\frac{2}{3}\%$
3. $87\frac{1}{2}\%$
4. $62\frac{1}{2}\%$
5. $11\frac{1}{9}\%$
6. $83\frac{1}{3}\%$

Page 47: Find the Percent

1. $12\frac{1}{2} = n$
   $12\frac{1}{2}\% = \frac{1}{8}$
2. $37\frac{1}{2}\% = \frac{3}{8}$
3. $66\frac{2}{3}\% = \frac{2}{3}$
4. $33\frac{1}{3}\% = \frac{1}{3}$

Page 48: Picture These Percents

1. $\frac{7}{10} = \frac{70}{100} = 70\%$
2. $\frac{3}{8} = \frac{37\frac{1}{2}}{100} = 37\frac{1}{2}\%$
3. $\frac{10}{20} = \frac{50}{100} = 50\%$
4. $\frac{3}{4} = \frac{75}{100} = 75\%$
5. $\frac{4}{20} = \frac{20}{100} = 20\%$
6. $\frac{3}{9} = \frac{33\frac{1}{3}}{100} = 33\frac{1}{3}\%$

Page 49: Mixed Review

	Fraction	Decimal (hundredths)	Percent
1.	$\frac{3}{5}$	.60	60%
2.	$\frac{3}{4}$	.75	75%
3.	$\frac{1}{3}$	$.33\frac{1}{3}$	$33\frac{1}{3}\%$
4.	$\frac{1}{5}$	.20	20%
5.	$\frac{1}{2}$	.50	50%
6.	$\frac{7}{10}$	.70	70%
7.	$\frac{1}{8}$	$.12\frac{1}{2}$	$12\frac{1}{2}\%$
8.	$\frac{1}{4}$	.25	25%

CHAPTER 6: MORE FRACTIONS AND PERCENTS

Page 50: Change Fractions to Percents

1. $\frac{3}{4} = \frac{75}{100}$ 75 hundredths = 75%
2. $\frac{1}{3} = \frac{33\frac{1}{3}}{100}$ $33\frac{1}{3}$ hundredths = $33\frac{1}{3}\%$
3. $\frac{4}{5} = \frac{80}{100}$ 80 hundredths = 80%
4. $\frac{2}{25} = \frac{8}{100}$ 8 hundredths = 8%
5. $\frac{3}{8} = \frac{37\frac{1}{2}}{100}$ $37\frac{1}{2}$ hundredths = $37\frac{1}{2}\%$
6. $\frac{9}{10} = \frac{90}{100}$ 90 hundredths = 90%
7. $\frac{5}{6} = \frac{83\frac{1}{3}}{100}$ $83\frac{1}{3}$ hundredths = $83\frac{1}{3}\%$
8. $\frac{5}{8} = \frac{62\frac{1}{2}}{100}$ $62\frac{1}{2}$ hundredths = $62\frac{1}{2}\%$
9. $\frac{2}{3} = \frac{66\frac{2}{3}}{100}$ $66\frac{2}{3}$ hundredths = $66\frac{2}{3}\%$
10. $\frac{1}{6} = \frac{16\frac{2}{3}}{100}$ $16\frac{2}{3}$ hundredths = $16\frac{2}{3}\%$

Page 51: Using Division

	Divide	Change to a Percent
1.	$10\overline{)3.00}$ = .30	.30 = 30%
2.	$10\overline{)9.00}$ = .90	.90 = 90%
3.	$4\overline{)3.00}$ = .75	.75 = 75%
4.	$5\overline{)3.00}$ = .60	.60 = 60%
5.	$2\overline{)1.00}$ = .50	.50 = 50%

Page 52: Working with Remainders

	Divide	Change to a Percent
1.	$8\overline{)7.00}$ = $.87\frac{4}{8} = .87\frac{1}{2}$	$.87\frac{1}{2} = 87\frac{1}{2}\%$
2.	$8\overline{)3.00}$ = $.37\frac{4}{8} = .37\frac{1}{2}$	$.37\frac{1}{2} = 37\frac{1}{2}\%$
3.	$8\overline{)5.00}$ = $.62\frac{4}{8} = .62\frac{1}{2}$	$.62\frac{1}{2} = 62\frac{1}{2}\%$
4.	$9\overline{)1.00}$ = $.11\frac{1}{9} = .11\frac{1}{9}$	$.11\frac{1}{9} = 11\frac{1}{9}\%$
5.	$3\overline{)2.00}$ = $.66\frac{2}{3}$	$.66\frac{2}{3} = 66\frac{2}{3}\%$
6.	$6\overline{)5.00}$ = $.83\frac{2}{6} = .83\frac{1}{3}$	$.83\frac{1}{3} = 83\frac{1}{3}\%$

Page 53: Mixed Numbers to Percents

	Improper Fraction	Think Division	Write Percent
1.	$\frac{15}{4}$	$4\overline{)15.00}$ = 3.75	375%

Page 53: Mixed Numbers to Percents (continued)

	Improper Fraction	Think Division	Write Percent
2.	$\frac{7}{3}$	$3\overline{)7.00}$ = $2.33\frac{1}{3}$	$233\frac{1}{3}\%$
3.	$\frac{9}{2}$	$2\overline{)9.00}$ = 4.50	450%
4.	$\frac{23}{6}$	$6\overline{)23.00}$ = $3.83\frac{2}{6} = 3.83\frac{1}{3}$	$383\frac{1}{3}\%$
5.	$\frac{23}{8}$	$8\overline{)23.00}$ = $2.87\frac{4}{8} = 2.87\frac{1}{2}$	$287\frac{1}{2}\%$
6.	$\frac{25}{4}$	$4\overline{)25.00}$ = 6.25	625%

Page 54: Use the Symbols

1. $>$
2. $<$
3. $<$
4. $=$
5. $>$
6. $<$
7. $>$
8. $<$
9. $>$
10. $>$
11. $<$
12. $>$
13. $<$
14. $>$
15. $>$
16. $=$

CHAPTER 7: PERCENT PROBLEM SOLVING

Page 55: Understanding Percents

1. $\frac{2}{20} = \frac{1}{10}$ .10 10%
2. $\frac{6}{24} = \frac{1}{4}$ .25 25%
3. $\frac{9}{10}$ .90 90%
4. $\frac{3}{4}$ .75 75%
5. $\frac{8}{10} = \frac{4}{5}$ .80 80%
6. $\frac{10}{30} = \frac{1}{3}$ $.33\frac{1}{3}$ $33\frac{1}{3}\%$

Page 56: Explain the Meaning

Answers should be similar to these:

1. All students were present.
2. $\frac{1}{2}$ off the original price
3. For every $1.00 on the bill, Allen paid 15¢.
4. $\frac{1}{3}$ off the original price
5. 35 parts out of 100 are wool and 65 parts out of 100 are cotton.
6. 9 out of 10 students passed the test.
7. Food that cost $1 last year costs $1.10 this year.
8. For every dollar's worth of goods sold, the commission is 20¢.

Page 57: Percents of a Circle

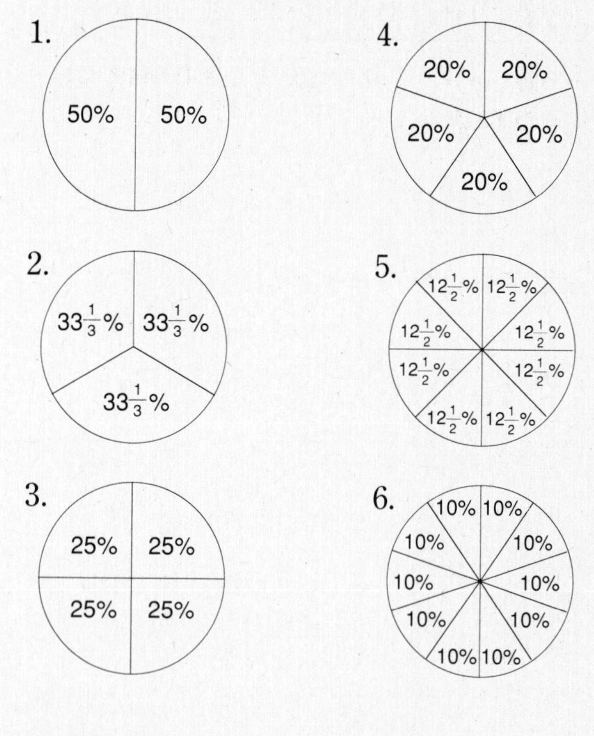

Page 58: Circle Graphs

1. 20%
2. 30%
3. 25%
4. 25%
5. delivering papers
6. 35%
7. food
8. rent
9. utilities
10. utilities

Page 59: Practice Your Skills

1. (X X X) X X
   (X X X) X X
2. ● ● ○ ○ ○
3. 75%
4. $33\frac{1}{3}\%$
5. 30%
6. 70%
7. 20%
8. 80%
9. $66\frac{2}{3}\%$
10. $33\frac{1}{3}\%$

Page 60: Percent Review

1. 25%
2. 40%
3. $66\frac{2}{3}\%$
4. ■ ■ ■ ■ □
5. ■ □ □ □
   ■ □ □ □
   ■ □ □ □

	Fraction (Simplified)	Decimal (Hundredths)	Percent
6.	$\frac{1}{2}$	.50	50%
7.	$\frac{1}{4}$	.25	25%
8.	$\frac{3}{10}$	.30	30%
9.	$\frac{3}{4}$	.75	75%
10.	$\frac{1}{5}$	.20	20%
11.	$\frac{3}{8}$	$.37\frac{1}{2}$	$37\frac{1}{2}\%$
12.	$\frac{1}{20}$	.05	5%
13.	$\frac{1}{100}$	.01	1%
14.	$\frac{1}{8}$	$.12\frac{1}{2}$	$12\frac{1}{2}\%$
15.	$\frac{1}{3}$	$.33\frac{1}{3}$	$33\frac{1}{3}\%$

CHAPTER 1: PERCENT PROBLEMS

Page 1: Percent Applications

1. \$5.00 2. \$1.50 3. 30%

Page 2: Percent Readiness

	Part Circled	Total—100% of Xs	Percent
1.	2	4	50%
2.	9	12	75%
3.	8	20	40%
4.	3	9	$33\frac{1}{3}$%
5.	2	10	20%
6.	5	8	$62\frac{1}{2}$%

Page 3: Understanding Number Sentences

	Diagram	Part Circled	Total—100% of Xs	Percent
1.	(X) X X X (X) X X X	2	8	25%
2.	(X X X) X	3	4	75%
3.	(X) X X X X (X) X X X X (X) X X X X	3	15	20%
4.	(X) X X	1	3	$33\frac{1}{3}$%
5.	(X) X (X) X (X) X X X X X	3	10	30%
6.	(X) X X X (X) X X X X X X X X X X X	2	16	$12\frac{1}{2}$%

Page 4: Showing Relationships

1. 10
2. 4
3. 40%
4. (X X X X) X
   X X X X X

	Total	Part	Percent
5.	70	7	10%
6.	50	30	60%
7.	10	3	30%
8.	9	6	$66\frac{2}{3}$%
9.	300	100	$33\frac{1}{3}$%
10.	90	45	50%
11.	12	10	$83\frac{1}{3}$%
12.	60	42	70%

Page 5: Identify the Part, Total, and Percent

	Part	Total	Percent
1.	16	80	20%
2.	21	30	70%
3.	5.7	38	15%
4.	192	96	200%
5.	8.04	67	12%
6.	63	70	90%
7.	267	89	300%
8.	28	70	40%

CHAPTER 2: FIND THE PART

Page 6: Find the Percent of a Number

1. $.30 \times 70 = n$
   $21 = n$

2. $.80 \times 50 = n$
   $40 = n$

Page 7: Change Percents to Decimals

1. $.19 \times 17 = n$
   $3.23 = n$
2. $.95 \times 960 = n$
   $912 = n$
3. $.08 \times 23 = n$
   $1.84 = n$
4. $.82 \times 194 = n$
   $159.08 = n$
5. $.48 \times 39 = n$
   $18.72 = n$
6. $.15 \times 598 = n$
   $89.7 = n$

Page 8: Small and Large Percents of a Number

A. $1.27 \times 89 = n$
   $113.03 = n$

B. $.004 \times 89 = n$
   $.356 = n$

1. $2 \times 348 = n$
   $696 = n$
2. $1.12 \times 95 = n$
   $106.4 = n$
3. $3.45 \times 70 = n$
   $241.5 = n$
4. $.008 \times 32 = n$
   $.256 = n$
5. $.003 \times 285 = n$
   $.855 = n$
6. $.025 \times 321 = n$
   $8.025 = n$

Page 9: Change Percents to Fractions

1. $\frac{1}{2} \times 80 = n$
   $40 = n$
2. $\frac{1}{5} \times 40 = n$
   $8 = n$
3. $\frac{1}{10} \times 60 = n$
   $6 = n$
4. $\frac{1}{4} \times 32 = n$
   $8 = n$

Page 10: Using Fractions

1. $\frac{1}{2} \times 20 = n$
   $10 = n$
2. $\frac{1}{10} \times 90 = n$
   $9 = n$
3. $\frac{1}{5} \times 50 = n$
   $10 = n$
4. $\frac{3}{4} \times 80 = n$
   $60 = n$
5. $\frac{1}{4} \times 120 = n$
   $30 = n$
6. $\frac{1}{2} \times 180 = n$
   $90 = n$

Page 11: Using Common Equivalents

1. $\frac{1}{4} \times \frac{24}{1} = n$
   $6 = n$
2. $\frac{1}{10} \times \$1,450 = n$
   $\$145 = n$
3. $\frac{1}{3} \times 45 = n$
   $15 = n$
4. $\frac{1}{5} \times 25 = n$
   $5 = n$
5. $\frac{1}{8} \times 720 = n$
   $90 = n$
6. $\frac{3}{4} \times 88 = n$
   $66 = n$

Page 12: Mixed Practice

Number sentences may vary.

1. $\frac{1}{2} \times 90 = n$
   $45 = n$
2. $.39 \times 145 = n$
   $56.55 = n$
3. $\frac{1}{4} \times 36 = n$
   $9 = n$
4. $.05 \times 4,508 = n$
   $225.4 = n$
5. $\frac{1}{3} \times 15 = n$
   $5 = n$
6. $1.75 \times 36 = n$
   $63 = n$
7. $.003 \times 910 = n$
   $2.73 = n$
8. $\frac{1}{10} \times 300 = n$
   $30 = n$
9. $.80 \times \$58.65 = n$
   $\$46.92 = n$
10. $\frac{1}{5} \times 35 = n$
    $7 = n$

Page 13: Write a Number Sentence

1. a) 15%
   b) $495
   c) *n*
2. 15% of $495 = *n*
   $.15 \times \$495 = n$
   $\$74.25 = n$
3. a) 25%
   b) $840
   c) *n*
4. 25% of $840 = *n*
   $\frac{1}{4} \times \$840 = n$
   $\$210 = n$

Page 14: Solve the Word Problems

1. 25% of \$32 = n
   \$8 = n

2. 5% of \$94.60 = n
   \$4.73 = n

3. 8% of 600 = n
   48 = n

4. 75% of 140 = n
   105 = n

5. 12% of \$347 = n
   \$41.64 = n

6. 30% of \$435 = n
   \$130.50 = n

7. 50% of \$300 = n
   \$150 = n

8. 10% of \$69,500 = n
   \$6,950 = n

CHAPTER 3: FIND THE PERCENT

Page 15: Compare the Numbers

	Picture	Fraction	Decimal	Percent
1.		$\frac{3}{12} = \frac{1}{4} = \frac{25}{100}$	.25	25%
2.		$\frac{4}{20} = \frac{20}{100}$	.20	20%
3.		$\frac{7}{10} = \frac{70}{100}$	.70	70%
4.		$\frac{4}{8} = \frac{1}{2} = \frac{50}{100}$	.50	50%
5.		$\frac{37}{50} = \frac{74}{100}$	.74	74%

Page 16: Find the Percent

1. $\frac{8}{10} = \frac{80}{100} = 80\%$
2. $\frac{7}{25} = \frac{28}{100} = 28\%$
3. $\frac{18}{20} = \frac{90}{100} = 90\%$
4. $\frac{3}{4} = \frac{75}{100} = 75\%$

Page 17: Simplify the Fraction

1. $\frac{5}{25} = \frac{20}{100} = 20\%$

Page 17: Simplify the Fraction (continued)

2. $\frac{15}{30} = \frac{1}{2} = \frac{50}{100} = 50\%$
3. $\frac{4}{40} = \frac{1}{10} = \frac{10}{100} = 10\%$
4. $\frac{24}{32} = \frac{3}{4} = \frac{75}{100} = 75\%$

Page 18: More Practice Finding the Percent

1. $\frac{8}{40} = \frac{1}{5} = \frac{20}{100} = 20\%$
2. $\frac{10}{40} = \frac{1}{4} = \frac{25}{100} = 25\%$
3. $\frac{3}{20} = \frac{15}{100} = 15\%$
4. $\frac{3}{10} = \frac{30}{100} = 30\%$
5. $\frac{4}{50} = \frac{8}{100} = 8\%$
6. $\frac{32}{40} = \frac{4}{5} = \frac{80}{100} = 80\%$
7. $\frac{12}{16} = \frac{3}{4} = \frac{75}{100} = 75\%$
8. $\frac{7}{70} = \frac{1}{10} = \frac{10}{100} = 10\%$

Page 19: Use Division

1. $\frac{5}{8} = 62\frac{1}{2}\%$
2. $\frac{15}{48} = \frac{5}{16} = 31\frac{1}{4}\%$
3. $\frac{5}{6} = 83\frac{1}{3}\%$
4. $\frac{30}{32} = \frac{15}{16} = 93\frac{3}{4}\%$

Page 20: Changing Fractions to Decimals

1. $\frac{3}{11} = 27\frac{3}{11}\%$
2. $\frac{7}{21} = \frac{1}{3} = 33\frac{1}{3}\%$
3. $\frac{3}{21} = \frac{1}{7} = 14\frac{2}{7}\%$
4. $\frac{21}{24} = \frac{7}{8} = 87\frac{1}{2}\%$
5. $\frac{2}{18} = \frac{1}{9} = 11\frac{1}{9}\%$

Page 20: Changing Fractions to Decimals (continued)

6. $\frac{14}{21} = \frac{2}{3} = 66\frac{2}{3}\%$

Page 21: What Is the Percent?

1. $\frac{6}{24} = 25\%$
2. $\frac{7}{35} = 20\%$
3. $\frac{7}{21} = 33\frac{1}{3}\%$
4. $\frac{25}{50} = 50\%$
5. $\frac{16}{24} = 66\frac{2}{3}\%$
6. $\frac{15}{40} = 37\frac{1}{2}\%$
7. $\frac{5}{7} = 71\frac{3}{7}\%$
8. $\frac{32}{40} = 80\%$

Page 22: Fractions and Percents Greater Than 100%

	Fraction	Percent
1.	$\frac{13}{10} = \frac{130}{100}$	130%
2.	$\frac{9}{5} = \frac{180}{100}$	180%
3.	$\frac{6}{4} = \frac{150}{100}$	150%
4.	$\frac{7}{2} = \frac{350}{100}$	350%
5.	$\frac{32}{20} = \frac{160}{100}$	160%

Page 23: Fractions Larger Than 1

1. $\frac{5}{4} = \frac{125}{100} = 125\%$
2. $\frac{15}{10} = \frac{150}{100} = 150\%$
3. $\frac{9}{2} = \frac{450}{100} = 450\%$
4. $\frac{50}{20} = \frac{250}{100} = 250\%$
5. $\frac{21}{6} = \frac{7}{2} = \frac{350}{100} = 350\%$
6. $\frac{35}{20} = \frac{175}{100} = 175\%$
7. $\frac{40}{25} = \frac{160}{100} = 160\%$
8. $\frac{22}{4} = \frac{550}{100} = 550\%$

Page 24: Decimals and Percents Greater Than 100%

1. $\frac{9}{2} = 450\%$
2. $\frac{10}{3} = 333\frac{1}{3}\%$
3. $\frac{8}{5} = 160\%$
4. $\frac{15}{7} = 214\frac{2}{7}\%$
5. $\frac{16}{6} = 266\frac{2}{3}\%$
6. $\frac{13}{4} = 325\%$

Page 25: Practice Your Skills

1. $\frac{9}{20} = 45\%$
2. $\frac{17}{85} = 20\%$
3. $\frac{27}{9} = 300\%$
4. $\frac{5}{2} = 250\%$
5. $\frac{15}{60} = 25\%$
6. $\frac{8}{7} = 114\frac{2}{7}\%$
7. $\frac{22}{5} = 440\%$
8. $\frac{20}{40} = 50\%$
9. $\frac{18}{4} = 450\%$
10. $\frac{12}{18} = 66\frac{2}{3}\%$

Page 26: Solve the Word Problems

1. a) $n\%$
   b) 20
   c) 15
   d) $\frac{15}{20} = 75\%$
2. a) $n\%$
   b) 30
   c) 6
   d) $\frac{6}{30} = 20\%$
3. a) $n\%$
   b) 16
   c) 4
   d) $\frac{4}{16} = 25\%$
4. a) $n\%$
   b) 25
   c) 8
   d) $\frac{8}{25} = 32\%$

Page 27: More Word Problems

1. $n\%$ of 45 = 15
   $\frac{15}{45} = 33\frac{1}{3}\%$
2. $n\%$ of 24 = 6
   $\frac{6}{24} = 25\%$
3. $n\%$ of 400 = 40
   $\frac{40}{400} = 10\%$
4. $n\%$ of \$60 = \$30
   $\frac{30}{60} = 50\%$
5. $n\%$ of 6 = 4
   $\frac{4}{6} = 66\frac{2}{3}\%$
6. $n\%$ of \$450 = \$90
   $\frac{90}{450} = 20\%$

Page 27: More Word Problems (continued)

7. $n\%$ of 30 = 10

   $\frac{10}{30} = 33\frac{1}{3}\%$

8. $n\% \times 15 = 12$

   $\frac{12}{15} = 80\%$

CHAPTER 4: FIND THE TOTAL

Page 28: What Is the Total?

1. $.30 \times n = 21$
   $n = 70$
   30% of 70 is 21.
2. $.40 \times n = 32$
   $n = 80$
   40% of 80 is 32.
3. $.80 \times n = 16$
   $n = 20$
   80% of 20 is 16.
4. $.68 \times n = 153$
   $n = 225$
   68% of 225 is 153.

Page 29: Find the Total

1. $.75 \times n = 225$
   $n = 300$
2. $.04 \times n = 34$
   $n = 850$
3. $.25 \times n = 15$
   $n = 60$
4. $.15 \times n = \$4.65$
   $n = \$31$

Page 30: Find the Total When the Part Is Given

1. $.03 \times n = 4.23$
   $n = 141$
2. $.60 \times n = 42$
   $n = 70$
3. $.06 \times n = 18$
   $n = 300$
4. $.10 \times n = 6$
   $n = 60$
5. $.25 \times n = 40$
   $n = 160$
6. $.80 \times n = 4$
   $n = 5$
7. $.15 \times n = 30$
   $n = 200$
8. $.90 \times n = 45$
   $n = 50$

Page 31: Percent Problem Solving

1. a) 20%
   b) n
   c) $8.00
   d) 20% of n = $8.00
   e) $40.00

Page 31: Percent Problem Solving (continued)

2. a) 25%
   b) n
   c) $150.00
   d) 25% of n = $150.00
   e) $600.00

Page 32: Solve the Word Problems

1. 4% of n = $28.00
   $700.00
2. 15% of n = $1.80
   $12.00
3. 20% of n = $156.00
   $780.00
4. 10% of n = 3
   30
5. 40% of n = $200.00
   $500.00
6. 15% of n = $39.00
   $260.00
7. 25% of n = $3.75
   $15.00
8. 20% of n = 49
   245

CHAPTER 5: PERCENT PROBLEM SOLVING

Page 33: Identify the Facts

1. 75% of 12 = 9
2. 25% of 80 = 20
3. 4% of 75 = 3
4. 10% of 30 = 3
5. 30% of 6 = 1.8
6. 45% of 20 = 9

Page 34: Learn the Percent Circle

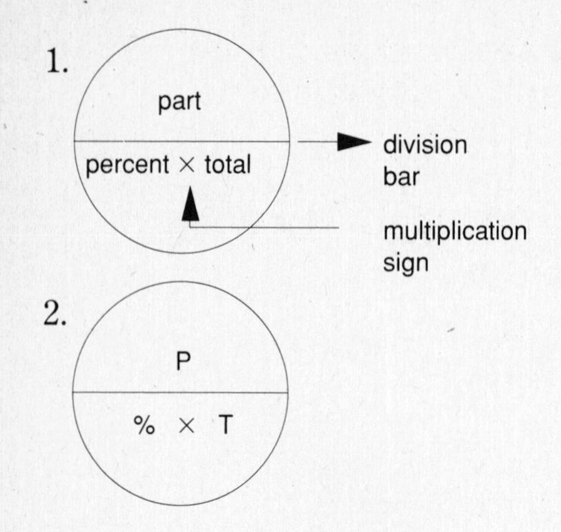

Page 35: Find the Part

1. a) P
   b) multiply
   c) 40% × 90 = 36

2. 15% × 75 = 11.25

3. 25% × 80 = 20

4. 30% × \$45 = \$13.50

Page 36: Find the Percent

1. a) %
   b) divide
   c) $\frac{7}{28} = \frac{1}{4} = 25\%$

2. $\frac{13}{65} = \frac{1}{5} = 20\%$

3. $\frac{8}{80} = \frac{1}{10} = 10\%$

4. $\frac{7}{175} = \frac{1}{25} = 4\%$

Page 37: Find the Total

1. a) T
   b) divide
   c) $\frac{7}{.20}$ $.20\overline{)7.00}$ = 35.

   7 is 20% of 35

2. $\frac{16}{.40}$ $.40\overline{)16.00}$ = 40.

   16 is 40% of 40

Page 37: Find the Total (continued)

3. $\frac{35}{.70}$ $.70\overline{)35.00}$ = 50.

   35 is 70% of 50

4. $\frac{9}{.20}$ $.20\overline{)9.00}$ = 45.

   9 is 20% of 45

Page 38: Use the Circle

1. a) percent
   b) divide
   c) 16 ÷ 48 = n
   d) $n = 33\frac{1}{3}\%$

2. a) part
   b) multiply
   c) .75 × 60 = n
   d) n = 45

3. a) total
   b) divide
   c) 17 ÷ .05 = n
   d) n = 340

Page 39: Problem-Solving Readiness

1. $\frac{1}{3} \times 45 = n$

   15

2. 15 ÷ 45 = n

   $33\frac{1}{3}\%$

3. 12 ÷ .05 = n

   240

4. n = 8 ÷ 50

   16%

5. .065 × 35.78 = n

   2.3257

6. .35 × 75 = n

   26.25

7. n = 40 ÷ 16

   250%

8. .88 × 325 = n

   286

9. 1.75 × 78 = n

   136.5

10. 9.35 ÷ .055 = n

    170

Page 40: Identify the Facts

A. 60%
B. 300
C. 180
D. 60% of 300 = 180

Page 40: Identify the Facts (continued)

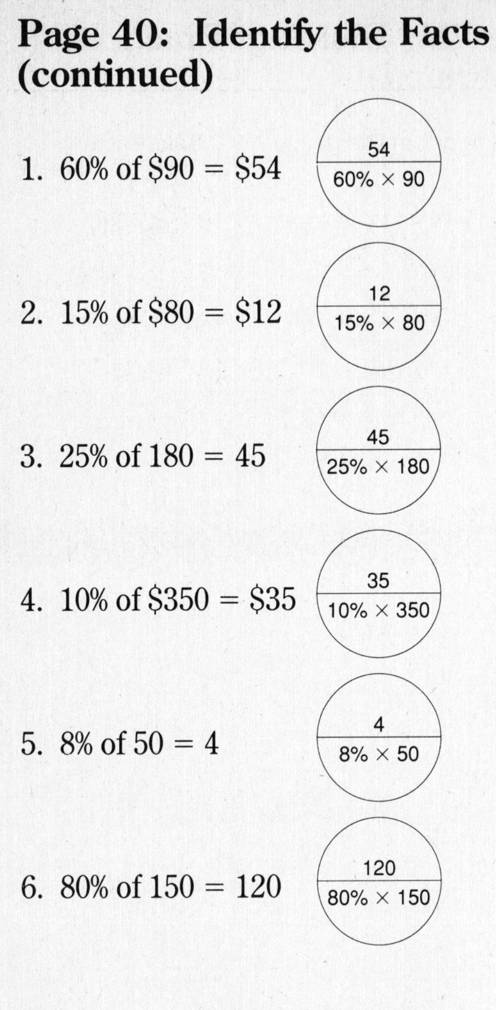

1. 60% of \$90 = \$54
2. 15% of \$80 = \$12
3. 25% of 180 = 45
4. 10% of \$350 = \$35
5. 8% of 50 = 4
6. 80% of 150 = 120

Page 41: Mixed Practice

1. 30% of \$20 = n; \$6.00
2. n% of \$40.00 = \$6.00; 15%
3. $33\frac{1}{3}$% of \$600 = n; \$200.00
4. n% of 5 = 3; 60%
5. 75% of 8,000 = n; 6,000
6. 18% of n = \$27.00; \$150.00
7. 10% of n = \$400.00; \$4,000
8. 4% of \$45.00 = n; \$1.80

Page 42: Problem Solving

1. 40% × n = \$34.00
   The original price was \$85.00.

2. 6% × \$12,540 = n
   The sales tax was \$752.40.

Page 42: Problem Solving (continued)

3. n% × \$4.50 = \$.90
   The tip was 20% of the fare.

4. 8% × n = \$2,240
   Her salary was \$28,000 before her raise.

5. $12\frac{1}{2}$% × 720 = n
   He sold 90 candy bars.

6. 5.5% × \$840 = n
   She earns \$46.20 in interest in one year.

7. 5% × n = \$136.00
   His total monthly earnings are \$2,720.

8. n% × 54 = 18
   $33\frac{1}{3}$% times at bat were hits.

Page 43: Think It Through

1. b
2. c
3. a
4. f
5. d
6. e

CHAPTER 6: LIFE-SKILLS MATH

Page 44: Round to the Nearest Cent

1. \$8.85
2. \$15.92
3. \$37.86
4. \$230.15
5. \$9.66
6. \$40.50
7. \$6.70
8. \$7.01
9. \$1.18
10. \$68.47

Page 45: Percent Off

1. 25% of \$88 = n
   \$22 = n
   \$22

2. 10% of \$399.95 = n
   \$39.995 = n
   \$40.00

Page 45: Percent Off (continued)

3. 30% of \$125 = n
   \$37.50 = n
   \$37.50

4. 40% of \$35.50 = n
   \$14.20 = n
   \$14.20

5. 15% of \$899.90 = n
   \$134.985 = n
   \$134.99

6. 50% of \$36.50 = n
   \$18.25 = n
   \$18.25

Page 46: Discounts

	Amount of Discount	Sale Price
1.	a) \$17.80	b) \$71.20
2.	a) \$5.10	b) \$10.20
3.	a) \$8.00	b) \$24.00
4.	a) \$62.80	b) \$62.80
5.	a) \$70.20	b) \$163.80

Page 47: Find the Sale Price

1. a) \$3.20
   b) \$12.79

2. a) \$124.75
   b) \$374.25

3. a) \$59.93
   b) \$39.96

4. a) \$3.89
   b) \$22.06

Page 48: Discount Practice

1. a) \$11.29
   b) \$33.87

2. a) \$45.00
   b) \$180.00

3. a) \$10.53
   b) \$24.57

4. a) \$698
   b) \$13,262

Page 48: Discount Practice (continued)

	Amount of Discount	Sale Price
5.	a) \$1.15	b) \$1.15
6.	a) \$11.84	b) \$47.36
7.	a) \$67.25	b) \$201.75
8.	a) \$.96	b) \$5.44
9.	a) \$3.38	b) \$13.50
10.	a) \$.26	b) \$2.34

Page 49: Sales Tax

Sales Tax

1. \$.52
2. \$1.27
3. \$.35
4. \$27.28
5. \$11.51

Page 50: Find the Total Price

1. a) 12.95
   b) 19.97
   c) 32.92
   d) 1.98
   e) \$34.90

2. a) 16.25
   b) 35.00
   c) 51.25
   d) 2.56
   e) \$53.81

3. a) 225.00
   b) 225.00
   c) 9.00
   d) \$234.00

4. a) 19.97
   b) 35.00
   c) 12.95
   d) 67.92
   e) 3.74
   f) \$71.66

Page 51: Sales Tax Practice

1. a) \$1.08
   b) \$19.03

2. a) \$.36
   b) \$9.35

3. a) \$6.29
   b) \$132.08

4. a) \$3.30
   b) \$69.28

Page 51: Sales Tax Practice (continued)

	Sales Tax	Total Purchase Price
5.	a) $2.93	b) $51.72
6.	a) $.24	b) $6.12
7.	a) $.68	b) $12.81
8.	a) $4.65	b) $71.04
9.	a) $1.20	b) $25.15
10.	a) $.25	b) $8.51

Page 52: Simple Interest in Savings

	Interest
1.	$7.50
2.	$48.00
3.	$11.00
4.	$22.80
5.	$27.00

Page 53: Borrowing Money and Paying Interest

	Interest	Amount to Be Repaid
1.	a) $63.00	b) $763.00
2.	a) $75.00	b) $325.00
3.	a) $1,200	b) $16,200
4.	a) $2,700	b) $10,200
5.	a) $560	b) $2,560

Page 54: Commission

1. 50% of $250 = *n*; $125.00
2. 15% of $82 = *n*; $12.30
3. 6% of $13,580 = *n*; $814.80
4. 20% of $5,960 = *n*; $1,192

Page 55: Commission Applications

1. 8% of $93,600 = *n*
   .08 × $93,600 = *n*
   $7,488 = *n*

Page 55: Commission Applications (continued)

2. *n*% of $1,590 = $79.50
   n = $79.50 ÷ $1,590
   n = 5%

3. *n*% of $350 = $175
   n = $175 ÷ $350
   n = 50%

4. 5% of *n* = $3,270
   n = $3,270 ÷ .05
   n = $65,400

5. 10% of $2,850 = *n*
   .10 × $2,850 = *n*
   $285 = *n*

6. 25% of $6,480 = *n*
   .25 × $6,480 = *n*
   $1,620 = *n*

7. 8% of $10,942 = *n*
   .08 × $10,942 = *n*
   $875.36 = *n*

8. 20% of $850 = *n*
   .20 × $850 = *n*
   $170 = *n*

Page 56: Practice with Commissions

1. 6% of $75,950 = *n*
   .06 × $75,950 = *n*
   $4,557 = *n*

2. 25% of *n* = $240
   n = $240 ÷ .25
   n = $960

3. *n*% of $1,200 = $240
   n = $240 ÷ $1,200
   n = 20%

4. 6% of $95,292 = *n*
   .06 × $95,292 = *n*
   $5,717.52 = *n*

Page 56: Practice with Commissions (continued)

	Business	Rate	Total Sales	Commission
5.	Real Estate	4%	$75,300	$3,012
6.	Tile and Carpeting	20%	$475	$95.00
7.	Appliances	8%	$300.00	$24.00
8.	Insurance	5%	$890.00	$44.50
9.	Automobile	10%	$15,700	$1,570
10.	Clothing	12%	$800	$96.00

Page 57: Percents and Budgets

1. $220
2. $1,100
3. $550
4. $300
5. $180
6. $480
7. $240

Page 58: Percent of Increase

1. 25%
2. 30%
3. 10%
4. 5%

Page 59: Percent of Decrease

1. 10%
2. 30%
3. 20%
4. 40%

Page 60: Percent Application Review

1. $4.97
2. 30%
3. $24
4. a) 32
   b) 15%
   c) 120
5. a) $3.99
   b) $83.79
6. $21.60
7. $1,344
8. $3,032
9. 25%
10. a) $5.78
    b) $32.74